TOXICOLOGY

Michael A. Kamrin

— **A Primer on Toxicology Principles and Applications**

- **Indoor & Outdoor Air**
- **Drinking Water**
- **Food**
- **Workplace Environment**

LEWIS PUBLISHERS

Library of Congress Cataloging-in-Publication Data

Kamrin, Michael A.
   Toxicology: a primer on toxicology principles and
applications
         p.   cm.
   Bibliography: p.
   Includes index.
   ISBN 0-87371-133-5
   1. Toxicology. I. Title.
   [DNLM: 1. Toxicology.   QV 600 K15t]
RA1211.K26 1988
615.9—dc19
DNLM/DLC   87-32492

Third Printing 1989

Second Printing 1988

LEWIS PUBLISHERS, INC.
121 South Main Street, Chelsea, Michigan 48118

PRINTED IN THE UNITED STATES OF AMERICA

Preface

As the use of synthetic chemicals increased dramatically after World War II, concern about the possible adverse health and environmental effects of such use grew. Chemicals with strange-sounding names, such as dioxin and toxaphene, and those known by initials, such as DDT, PCB, and EDB, became familiar to everyone. Unfortunately, the rate at which chemicals were developed and distributed was not matched by the rate of increased scientific or public understanding. As a result, our society has faced and continues to face a number of dilemmas in trying to deal effectively with toxic substances. The purpose of this book is to familiarize the reader with our current knowledge of toxicology and with how our society uses this understanding in managing toxic substances. To accomplish this, the book opens with a discussion of the general principles of toxicology and how these general principles are applied in assessing the acute, subacute and chronic effects of chemicals. Both qualitative and quantitative measures of toxicity are addressed and the protocols for the various tests will be described.

In the course of this discussion of general principles, emphasis will be placed on both the strengths and limitations of the techniques which are in current use. The uncertainties that result from the limits of our understanding will also be emphasized. Finally, the influence of societal attitudes on the testing that is performed will be discussed.

This treatment of general principles is followed by a discussion of risk assessment and risk management. Risk assessment deals with both the toxic potential of a chemical and the exposure likely to occur in specific situations. Risk management includes the steps taken to reduce or eliminate risks identified in the assessment process. The components of risk assessment are described in relation to a variety of possible toxic effects and exposure scenarios. Alternative risk management strategies are presented, along with the approaches most commonly employed in the United States.

The last part of the book provides a description and analysis of four case studies, each dealing with a different substance. These include the food additives, cyclamate and saccharin; the fire retardant material, asbestos; the widely-used formaldehyde; and the petroleum product, benzene. The basic toxicological information regarding each, and the different regulations to which they have been subject, provide the basis for exploring how the principles introduced in the first part of the book are applied in practice.

The summary draws together similarities and differences evident in the case studies and uses these to illustrate the factors which influence the way toxic substances are managed in the United States and elsewhere. The complexities in trying to manage substances that are ubiquitous and closely tied to our way of living can be readily appreciated by the comparison. In addition, the serious difficulties facing citizens who must make choices on an individual basis become evident.

Overall, this volume attempts to help the reader put toxic substances in perspective and to appreciate the inherent uncertainties in trying to answer many questions about toxic effects of chemicals. In understanding these limitations, action can be taken that is more realistic and thus more likely to be successful. Knowledge is critical to informed choice and to a more enlightened era in dealing with toxic substances.

Michael A. Kamrin, born and raised in Brooklyn, New York, earned a BA in Chemistry from Cornell University and an MS and PhD in Biophysical Chemistry from Yale University. He pursued his research interest in energy transduction processes during postdoctoral appointments at Oak Ridge National Laboratory and Stanford University. His particular interest was the transformation of light energy into chemical energy during photosynthesis. Following this research, he joined Michigan State University in the Department of Natural Science, a department which provides general education in science for nonscience students. While examining the role of science in public policy, he became interested in the specialized area of risk assessment. Dr. Kamrin developed this interest during 15 years of teaching at MSU and, in 1980, gained a different view of societal applications of risk assessment as visiting scientist in the Michigan Legislative Office of Science Advisor. After a year's service, he returned to MSU as a professor in the newly formed Center for Environmental Toxicology, a position he still holds. His main role in the center is the translation of scientific concepts in environmental toxicology into language which can be understood by a variety of audiences. This involves responses to public inquiries, workshops for concerned citizens, programs for extension agents, Lifelong Education courses, interactions with public agencies, materials for curriculum development, and publications for a variety of groups. In the past year, he has focused on hazardous waste issues and problems in risk communication.

Contents

List of Figures

List of Tables

1

Introduction

Toxicology, also known as the science of poisons, has a long tradition, although it was not considered an independent discipline until quite recently. There is evidence that early in the history of the human race people understood the potency of certain natural materials and used them in hunting and for their therapeutic values. Some familiar images from the past reinforce this belief. For example, the story of Socrates ingesting poison hemlock and the stereotype of the royal tasters are in this tradition. As time went on, and especially following the scientific revolution, the understanding of how chemicals affect the human body fell into the domain of pharmacology. Both beneficial and adverse effects were studied under this general heading, and so toxicology was but a branch or subset of pharmacology.

The rise of an independent science of toxicology is a twentieth century phenomenon. It appears to follow the growth of the chemical industry and the concomitant production of synthetic chemicals, compounds which led to somewhat different types of health problems than were previously faced with natural toxicants. During the late 1940s and the 1950s, attention was focused on air pollutants and the occurrence of generalized health problems in polluted areas, as well as a few very serious incidents which led to numerous fatalities. The 1960s were marked by attention to other products, such as DDT, and to the pollution of water from waste streams emitted by industry. In the 1970s and 1980s, the focus has shifted to a variety of other chemicals, such as PBBs and dioxins, and to the general problem of contamination of air, water, and soil by improperly contained wastes, especially in landfills.

1

The industrial toxicology laboratory, where chemicals could be tested before being produced in large quantities, dates back into the 1930s, albeit in a much less sophisticated form than is found today. The national recognition of the science of toxicology in the United States can probably be traced to the formation of the Society of Toxicology in 1961. This institutionalized the growing number of scientists who were involved in the testing of potential environmental pollutants and also a wide variety of other chemicals, including food additives and, of course, drugs. However, it was not until the mid-1970s and the 1980s that toxicology started to permeate academic institutions with the formation of departments having toxicology as part of their title. In some cases, it was as an addendum—Department of Pharmacology and Toxicology. In others, it was more specialized—Department of Environmental Toxicology. In addition, graduate programs in toxicology began and, very recently, in a handful of institutions, undergraduates were given the option of majoring in this field.

Thus, toxicology is a very young discipline and has a limited number of truly trained disciples. Many scientists who are called toxicologists are trained in other fields and have become toxicologists through experience grafted onto their previous education. There have been attempts to limit this type of self-appellation by the use of certification exams, but only a small percentage of current "toxicologists" are certified. At present, the demand for toxicologists much exceeds the supply, so that there is not significant pressure on most to be certified. This means that scientists with a wide variety of expertise and backgrounds may be perceived as equally expert by the public and have equally significant effects on the public image of this science as well as public policy in this area. The differing levels of expertise have contributed to the apparent conflict and uncertainty in the scientific community regarding the toxic effects of a number of chemicals.

Although toxicology seems to have many facets—such as environmental toxicology, food toxicology, and clinical toxicology—it is a discipline with basically one goal. That is the understanding of how chemicals can adversely affect living organisms. Although this goal can be stated rather simply, there is a lot more to achieving it. There are thousands of chemicals, both natural and synthetic, which abound in our environment, with a multitude of possible adverse effects. Beyond these difficulties is the fundamental problem that our understanding of the normal functioning of living organisms is lacking in many ways. Thus, there are often questions as to whether or not an adverse effect has occurred, even if some biological change can be detected.

On first glance, this appraisal might lead to a rather pessimistic view of toxicology. Perhaps, it might appear to replace economics as the "dreary science." However, the history of science to date reveals that it is often possible to accomplish much with incomplete understanding *as long as* this state of ignorance is acknowledged. Medical practitioners have utilized many different kinds of chemicals without knowing exactly how they act—aspirin is a good example of this. Newton's ideas of physics, although later shown incorrect by Einstein, were good enough to significantly help the British fleet to defeat the Spanish Armada. The astronauts travel into space based on the knowledge that we have, but they also have built-in flexibility of action so they can respond to the unexpected. The limitations of our understanding are recognized and taken into account. Toxicology, like all science, can be applied to provide answers to certain questions we wish to address. However, these answers must be looked at carefully so we do not get a false sense of what we know and thus unrealistic expectations as to what particular actions will accomplish.

Unfortunately, we have often been taught that science produces certainty, and this is reinforced by the identification of science with technology. Scientists must know the truth if they can invent the telephone, TV, VCR, and disposable diaper. As a result, there is a certain impatience with scientists who express uncertainties of various sorts, and a tendency to believe other scientists who express their views without such reservations. Then, as often happens, the scientist who was more cautious turns out to have been correct, and some apparent finding is overturned. This sequence of events occurs with regularity, and tends to be interpreted not as an indictment of those who express certainty where none exists, but instead as an indictment of science and scientists in general. This type of situation is likely to occur even more often in a young science where less is known. Toxicology falls into this category.

Perhaps the first lesson from all this is that most definite answers in toxicology are likely to be oversimplifications, and thus generally inapplicable to most future situations. As a result, skepticism must be the rule rather than the exception. A corollary of this is that to ask for a definite and final answer is to ask to be deceived. Perhaps a few examples from an allied field, pharmacology, will make this point more clearly. Many people decided to take Laetrile as a cancer drug because it was advertised as a certain cure as opposed to other therapies which could only give certain odds for being successful. Subsequent events have shown it to be of little value. Others have taken huge amounts of vitamins because of the alleged certainty that they would prevent wrinkled skin, aging, impotence, and so on. Again, the reality was different from the promise. Medical quacks with sure cures are always around us and always seem to have many adherents. It is difficult to accept uncertainty, but it is often the most reasonable course of action.

In the field of toxicology, the certainty that most people seek is that a particular chemical is "safe," i.e., it will not cause adverse health effects in the human body. This goal is unachievable, as was recognized several centuries ago. There is no such thing as an absolutely safe chemical. All chemicals can cause toxic effects in large enough amounts. This includes the most vital of chemicals—the oxygen we breathe and the water we drink. When faced with this reality, most people look for a different certainty—a "safe" amount. They would like to know the exact level at which a chemical changes from a nontoxic to a toxic chemical. This, again, is not a scientifically realistic goal. People vary tremendously in their response to their environment, including the chemicals in it, so that what is "safe" for one person may not be "safe" for another.

At this point, you may be saying to yourself: But the government has set safe levels for many chemicals. This is indeed true. However, the rationale behind these standards is not entirely scientific in nature. Most people would like to entirely eliminate those chemicals which have been labeled as especially toxic, i.e., which can cause adverse effects in very small amounts. However, because of their ubiquitous distribution, the difficulty in detoxifying them, or their highly desirable benefits, they cannot be realistically eliminated. Thus, we must accept some level of these substances in our environment and, if we want to have an enforceable regulation, this level must be a definite number. It would be impractical to set a variable standard. Thus a single number becomes the dividing line between "safe" and "unsafe." The amount and quality of scientific evidence behind this number varies from case to case and often changes over time. It is not unusual for standards to be adjusted as scientific evidence changes. Vulnerability to change should be the first clue that the certainty of a "safe" level is not as definite as it appears.

Questions then arise: Where do these levels come from? How are they set? This varies from case to case. As will be explained in detail in the succeeding chapters, there are a number of factors involved. Some are scientific. Some are social. Some are economic. And some are historical in nature.

Different federal agencies may set different standards for the same chemical. Different countries often set different "safe" levels for the same substance. Particular compounds may be banned in one country but not in others. Although there is no one answer, it is important that everyone understand how each of the answers is determined so that they can act accordingly. It is clearly just as dangerous to act in the belief that knowledge is certain when it is not, as it is to act in the absence of any knowledge at all.

The science of toxicology is still maturing. A number of the early studies were performed by scientists who had a poor grasp of the basic tenets that should be applied. In retrospect, some current studies are also likely to be looked at less than kindly. Little solid information exists about the toxicity of a large number of chemicals, and the basic understanding of how most toxicants act is lacking. There are about 6,000,000 known chemicals, of which about 50,000 are in common use. Detailed chronic toxicity tests have been performed on only a few hundred of these. Even for those which have been tested, there are questions about the interpretation of the results obtained and serious reservations about the applicability of these laboratory test results to human populations in everyday situations.

Unfortunately, there is a continuing clamor on the part of the public to receive answers that cannot be provided with confidence, if at all. Thus toxicology must grow up in the limelight, and the hidden ''warts'' in most portrayals of the growth of scientific disciplines are painfully obvious. The fits and starts which have always characterized science, but which have been romanticized away in the historical accounts, are revealed to all in toxicology and seen as a fault of the toxicologists rather than the normal way that all science operates.

The focus of this book is to summarize where toxicology is today, and to reveal what basic principles are agreed on and how they are applied. It is also to show the difficulties which beset this science, and how these in turn are reflected in public attitudes and national policy. The aim is to provide a realistic appraisal of what toxicology is and what it can reveal.

2

Chemical Individuality and Toxicity

The toxic effects of a substance on a particular individual depend on both the chemical and the individual. However, the variability in the toxic potential of different compounds greatly exceeds the variability in toxic response from individual to individual. That is, if we expose a particular individual to a whole series of different chemicals, toxic effects would occur after administering minute (invisible) amounts of some substances, but only after administering huge quantities of others. The toxicity of one chemical can be millions, billions or even more times greater than the toxicity of another chemical; i.e., it will take millions or billions times as much of one chemical to cause the same effect as another. Human variability is not as great. If a particular chemical causes an effect in one individual when a particular amount is administered, it is not likely that an amount a billion times less will cause a toxic effect in another individual. The exact range of human variability is not well established, but it is probable that it is closer to a hundredfold than a billionfold.

2.1 TYPES OF CHEMICALS

Since the greatest source of variability in toxicity is the chemical involved, this will be explored first. Chemicals come in a wide variety of sizes and shapes. Naturally occurring compounds are found in the earth's crust, in the oceans, streams, and rivers, and in the atmosphere, and are produced by both plants and animals. Some of these chemicals have very sim-

7

ple structures, such as lead, while others, such as mycotoxins, are very complex molecules. Synthetic chemicals, which are manufactured for a variety of uses, may also be simple or complex. Examples of both natural and synthetic chemicals are given in Figure 2.1.

What often makes synthetic chemicals of particular concern is that they are different from naturally occurring ones and so present the living organism with a new kind of challenge. In some cases, these synthetic chemicals are very similar to natural ones and so deceive living organisms into treating them the same way. This may lead to an inappropriate response by the living organism and thus adverse effects. In other cases, these synthetic chemicals may overwhelm the organism's defense mechanisms or even avoid them because of their uniqueness. It should be emphasized that most chemicals, whether synthetic or natural, are not very toxic. It is not the source of the chemical but its characteristics that are important in determining toxicity.

2.2   SHAPES OF CHEMICALS

One of the most important characteristics of a chemical is its shape. It is thought that the living organism recognizes, and thus reacts to, most chemicals that enter the body as a result of their shape. Thus a protein that is present in the meat you eat is recognized by a molecule in your body, which attaches to it and starts a sequence of events which breaks it down into its components. These components are then recognized by other molecules, which take them and join them together to form the proteins needed by your body. Similarly, a foreign germ is recognized by defense molecules in your body by its shape, is attacked, and ultimately is removed from the body as a result of this recognition. These body recognition molecules respond to very subtle differences in shape. Two molecules which might look very similar when drawn on a piece of paper can have different three-dimensional structures, and one may lead to a toxic response in the body while the other will not, if both are given in the same amount. An

NATURALLY OCCURRING CHEMICALS

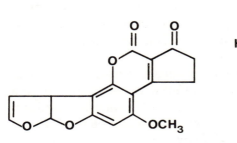

Aflatoxin B₁

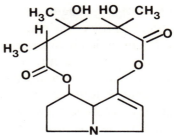

Monocrotaline

SYNTHETIC (MAN-MADE) CHEMICALS

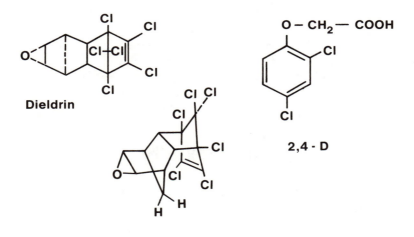

Dieldrin

2,4 - D

Figure 2.1   Structures of natural and synthetic chemicals found in the environment.

example of this type of subtle difference is given in Figure 2.2. In this case, there are exactly the same number of each type of atom—hydrogens, carbons, and chlorines—in each molecule, but the chlorines are placed differently in each. One arrangement leads to a fairly toxic molecule while the other leads to a relatively nontoxic one.

Sometimes even more subtle differences in structure result in great differences in biological impact. One branch of toxicology involves examining different but similar molecules to see if there are some patterns which can be detected. If toxicologists can establish a relationship between particular structural differences and differences in toxic effect, they will be able to predict the toxicity of a molecule before it is made. Unfortunately, this area of toxicology is quite new, and it is likely to be some time before such predictions will be possible.

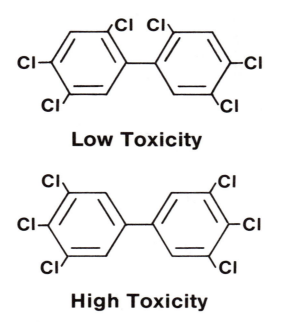

Figure 2.2 Example of structurally similar chemicals with different toxic potency.

2.3   POLARITIES OF CHEMICALS

A second important characteristic in determining the degree of toxicity a chemical will produce is how well it dissolves in different solvents. In particular, compounds are divided into those that are soluble in waterlike (or polar) solvents and those that are soluble in fat-(oily)-like (or nonpolar) solvents. This difference is very important in determining how easily a chemical can enter the body, how it is distributed inside the body, and how easily it can be excreted. An example of a polar compound (one which dissolves in polar solvents) is table salt. An example of a nonpolar compound (one which dissolves in nonpolar solvents) is DDT. Since the human body is most efficient at excreting polar compounds, there are chemicals in the body that are designed to convert foreign nonpolar compounds into polar variants so that they can be removed from the body. Unfortunately, they cannot alter the characteristics of all nonpolar compounds, and so some, like DDT, are difficult to excrete and can remain in the body for long periods of time. Briefly put, this one difference in character—polarity or nonpolarity—can have a very significant impact on the toxicity of a particular chemical.

2.4   DOSES OF CHEMICALS

As was discussed earlier, all chemicals, no matter what their characteristics, are toxic in large enough quantities. Thus the amount of a chemical a person is exposed to is crucial in determining the extent of toxicity that will occur. In attempting to place an exact number on the amount of a particular compound that is harmful, scientists recognize that the size of the organism has to be taken into account. It is unlikely, for example, that the same amount of a particular chemical that will cause toxic effects in a 1-pound rat will also cause toxicity in a 1-ton elephant.

Thus instead of using the amount that is administered or to which an organism is exposed, it is more realistic to use the amount per weight of organism. Thus it could be said that an amount of 1 ounce administered to a 1-pound rat is equivalent to 2000 ounces to a 2000 pound (1-ton) elephant. In each case, the amount per weight is the same: 1 ounce for each pound of animal.

This amount per weight is known as the dose. It is used to determine the amount of drug to prescribe for patients of differing weights and is used in toxicology to compare the toxicity of different chemicals in different animals. In general, metric units are used, and the toxicity is often expressed in milligrams per kilogram (written mg/kg). One milligram is one-millionth of 1 kilogram. It can be seen from these units that when scientists talk about a toxic chemical, they are talking about one that can cause adverse effects in an amount that is very small compared to the weight of the organism receiving this chemical.

2.5 DOSE-RESPONSE

Scientists often administer a wide variety of doses to experimental animals to determine what dose of a particular chemical causes which kinds of toxic effects. What they usually find is that a very small dose causes no observable effects, a higher dose some toxicity, still higher doses greater toxicity, and at a high enough dose the animal dies. This gradual increase in toxic effects is known as dose-response and is often presented graphically rather than in words. A typical dose-response curve is shown in Figure 2.3.

Now, if we assume that this dose-response curve represents what happens when chemical X is given to a rat, the next question is: What happens if chemical X is given to a mouse? Will the dose-response curve look the same? If not, how will it differ? The answer is that the dose-response curve will probably have a similar shape, but it will start rising at either a lower or higher dose than in the rat. There are a whole family of curves which describe what happens when chemical X is given to a number of different animals (see Figure 2.4). This shows that some animals are more sensitive to chemical X than others; i.e., it requires a smaller dose to produce the same toxic

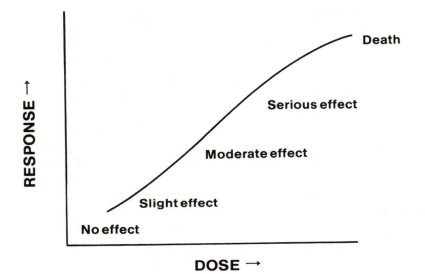

Figure 2.3 Dose-response relationship for a typical chemical.

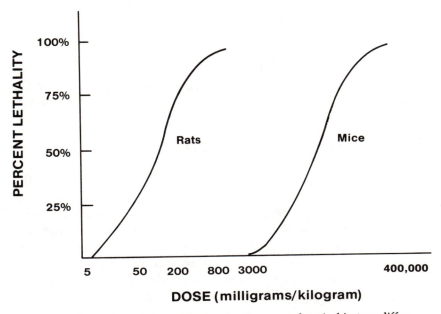

Figure 2.4 Variation of dose-response for the same chemical in two different species.

effect. The range of sensitivity can be very great. It can some-times take a thousand or ten thousand times as high a dose to produce toxicity in one animal as in another. This differ-ence is very important to consider when we are thinking about using toxicity data from animal experiments and applying them to human beings. Are human beings more sensitive, less sen-sitive, or just as sensitive to chemical X as a rat?

Instead of asking how animals vary in their response to chemical X, we can also ask how one particular type of animal responds to a wide variety of chemicals—chemical X, chemi-cal Y, chemical Z. If we take rats, for example, and administer these three chemicals we can produce a dose-response curve for each chemical. Again we will get a whole family of similar curves and may find that rats are most sensitive to chemical Z, for example, less sensitive to chemical X, and least sensi-tive to chemical Y. Expressed in a different way, chemical Z is the most toxic to rats, and chemical Y is the least toxic. It should be emphasized that this order of toxicity is specific to the rat and may or may not be the same for mice or guinea pigs or humans. However, it does provide a rough guide to the relative toxicity of various chemicals. The most toxic chem-icals are usually very toxic in a number of different types of animals, and the same holds for the least toxic. There may be some differences in order within these groups, but the general rankings will normally hold from species to species.

2.6 CATEGORIES OF TOXICITY

The above provides a qualitative or rough estimate of the relative toxicity of various chemicals. In order to make reprodu-cible quantitative determinations of toxicity, specific experi-mental protocols must be established and followed. There are two problems in establishing such protocols. One is that there are many different types of toxicity and thus the necessity for a number of possible protocols. The other is that establishing protocols involves trial and error, and thus is a dynamic process that has changed over time. Both of these problems are serious ones and contribute to a number of scientific questions and to the confusion that exists among the public about the meaning of toxicological information that is reported.

The first problem, multiple types of toxicity, has been approached by using a simple classification system based on the time needed to produce toxicity. Acute toxicity tests, which measure the earliest effects, are performed by administering a single dose (or a small number of doses) and assessing toxicity up to two weeks afterward. Subacute toxicity is measured using daily dosing during the first 10% (approximately) of the organism's lifetime and checking for effects during and at the end of this period. Tests of chronic toxicity are generally performed by daily dosing over a lifetime, and periodic examination of the exposed organisms during the course of dosing and at the end of the study period.

2.7   TYPES OF TOXIC EFFECTS

In addition to the timing of the appearance of the toxic effects, the type of toxicity can also vary tremendously. There may be local effects, such as skin irritation; and general effects, such as incoordination, behavioral changes, organ structure changes, and others. Since it would be very time consuming to set up tests for each effect during each time period, different approaches have been taken for the different categories of toxicity. In acute toxicity testing, death of the animal is the main criterion of toxicity. On the other hand, the presence of any type of toxic effect is an indicator of toxicity in subacute testing. The most common type of chronic toxicity testing, the carcinogenesis bioassay, utilizes one specific endpoint, cancer, as the criterion of toxicity. Thus acute, subacute, and chronic toxicity tests are performed using different endpoints and thus employing different protocols. The details of each of these types of toxicity assessment will be presented in Chapter 6.

2.8   TESTING PROTOCOLS

The second problem, the evolution of testing protocols, becomes important when comparisons are being made between toxicants that have been studied at different times or between studies of the same toxicant performed at different times. For example, early acute toxicity studies were sometimes done using a seven-day instead of a fourteen-day waiting period. This may have affected the results of the tests. Even if the general protocol remains unchanged over time, other important factors may differ from time to time. For instance, the diet may be improved from one period to the next. Experimental animals may change over time as breeding conditions lead to genetic alterations. Even the material to be tested may change over time as production processes are improved. For example, the degree of purity may become much greater. In addition, analytical techniques have improved quite dramatically over time, and earlier studies, which used relatively crude dosage measurements, have been replaced by tests that measure doses very exactly and allow the reproducible administration of very low dosages.

While these changes could be taken into account if the studies were repeated as protocols changed, limitations of time, money, and expertise make such repetitions unrealistic. Thus we are often faced with a collection of data about a chemical which represents studies made over several decades. It is not easy to decide how, or if, to use early data and what to do when crucial pieces of information about study protocols are missing. As a result, we must accept even greater uncertainties in toxicity assessment than might be expected from an examination of the current state of toxicity testing.

2.9   SUMMARY

In summary, the characteristics of each chemical greatly influence the type and severity of toxic effects they can produce. Two of the most important of these characteristics are the shape of the molecule and the ability of the chemical to dissolve in either polar or nonpolar solvents. In addition to the type of substance involved, the dose the organism receives is crucially important. Indeed, since all chemicals are toxic at some dose, the dose can be considered the most crucial factor in whether or not adverse effects will result from exposure to a chemical.

Unfortunately, the "toxic dose" is not an easy quantity to determine, since it varies from animal to animal. In addition, it depends on the particular definition of toxicity that is used in determining this dose. Thus toxicity is species-specific as well as chemical-specific.

3

Human Variability and Toxicity

A reading of any toxicology study, even those done on animals that are bred to be as similar as possible, reveals that different animals respond differently to the same administered dose. For example, in an acute study, a certain percentage of the animals die during the 14 days after dosing, rather than all or none. Similarly, in chronic assays, only a certain percentage develop cancer. Thus some animals must be more sensitive and others less sensitive to the same dose of a foreign chemical. Since this holds for a population bred to be similar, you would expect a much greater range of sensitivity in a wild animal population or a human population, and this is indeed the case.

3.1 GENETICS, AGE, AND SEX

What is the source of this variability? One very obvious reason that people react differently is that they are born differently; i.e., they have different genetic makeups. The chemicals produced in each individual are unique, not only in the exact forms of each chemical produced but also in the amount of each in the body. Thus each individual contains a different response system, which affects that person's ability to function under normal conditions and when challenged by toxic substances. This is obvious when we think about disease resistance. At one extreme we have the few cases of children who are born without disease resistance (the boy in the bubble) and at the other we have people who "have never been sick a day in their life." It is clear that genetic factors are critical in the first case and undoubtedly play a very significant role in the second.

Another source of variability in response is also genetic in nature, but it is a little different from what has just been discussed. It is the difference between males and females. As we all know, males and females do have different chemical compositions (most of which are related to the difference in reproductive and nurturing functions). However, as an unintended by-product of these differences, there is variation in the ability of males and females to deal with particular toxicants. It is not an across-the-board situation where one sex is more or less sensitive; it depends on the nature of the toxicant. As a simple illustration of how this might work, the different distribution of fat in male and female bodies leads to a different distribution of lipid-soluble toxicants in each. This in turn may lead to the occurrence of toxic effects at higher or lower doses. Such sex differences have been clearly demonstrated in experimental studies of responses to a variety of chemicals, both toxic and beneficial (drugs).

A third reason that one person's sensitivity differs from another's is their respective ages. Young children and very old people often are more sensitive than individuals in other age groups. This can be largely explained by the differences in body chemistry that occur with age. The very young often do not have fully developed body mechanisms that can effectively detoxify foreign chemicals. The very old may have lost this same ability, at least to some degree. In addition, certain types of toxic effects can only occur while development is still going on, so that adults would not be susceptible. For example, the full development of the brain takes about six or seven years after birth, and children who are younger than this are susceptible to toxic chemicals that can affect brain functions, such as learning ability or behavior.

3.2 HEALTH AND PREVIOUS EXPOSURES

A fourth source of variability in response to chemicals is the state of health of the individual at the time of exposure. This was clearly illustrated during the serious air pollution episodes of the 1940s and 1950s that were fatal to those individuals who already had severe respiratory problems, but produced much less serious effects in those with healthy lungs. It is not only the lungs which must be considered, but also a variety of other organs. The liver, for example, is the most important detoxifying organ in the body, and, if it is damaged, the individual will be much more susceptible to a wide variety of toxic agents. Even more subtle differences in health status, such as differences in blood pressure, may affect the degree of toxicity a particular chemical will produce in a specific individual.

A related source of variability is the previous chemical exposure of an individual before a particular encounter with a chemical. While this earlier episode may not have affected the health status of the individual, it may have caused chemical alterations that show up only when a new challenge is faced. For example, the initial exposure may have mobilized the body's defense mechanisms against such toxic substances by leading to the production of higher amounts of detoxifying substances than would normally be present. This might be thought an advantage in that it could reduce the toxic potential of these later exposures. While this is indeed the case in many instances, it may also be a disadvantage if the body's attempts at detoxification lead instead to the production of a chemical in forms more toxic than the chemical that has entered the body. Unfortunately, this situation is not that unlikely, especially with certain classes of synthetic chemicals. Thus previous exposure may lead to either increases or decreases in the severity of the toxic effects following later exposures to the same or different chemicals.

3.3 PSYCHOLOGICAL FACTORS

It should be pointed out that there is often a mental as well as a physical aspect to toxicology. It is well known in medicine that the administration of a placebo, an inert substance, can often lead to great improvement in the patient. This occurs even though the material administered does not have any direct effect on the problem. The feelings of the patients that the treatment will work has a great impact on whether or not they perceive that it does work. In a similar vein, the feeling that a material is harmful can lead some individuals to experience adverse effects even if there are no direct physical effects of that substance. This is a very serious problem when the effects of concern are subjective in nature, e.g., headache, irritability, etc., and thus are not amenable to external verification. Thus a sixth source of response variability from individual to individual should be added: the psychological state and makeup of the exposed individual.

Overall, there are many factors which can influence the severity of toxic effects (if any) that occur as a result of exposure to a chemical. These include genetic, historical, and psychological factors and together can result in a tremendous range of response from individual to individual. As a result, it is impossible to predict the response of any one individual to any particular exposure. This uncertainty is reflected in the probabilistic nature of most of the toxicity assessments that are made. They deal with a population, not an individual, and try to predict what percentage of people in that population will show a particular effect at a particular dose. This is the best that can be done, and, as will be clear later, even this limited type of prediction is filled with uncertainties.

4

The Human Body and Toxic Chemicals

The discussion of variability in the previous chapter presumes that each individual exposed to the same amount of chemical received the same dose of chemical at the site in the body where the toxic effect occurred, e.g., the lungs. However, it has been shown experimentally that this is not the case. One of the main factors that influences the dose at the site of action, or the "effective dose," is the route of exposure or how the individual was exposed. The three main routes of exposure are inhalation (breathing in the material), ingestion (eating or drinking the substance), and dermal contact (entry through the skin). From this simple analysis, it might appear that the most dangerous chemicals to the lungs are those that are inhaled; to the digestive system, those that are ingested; and to the skin, those that are touched. This fairly obvious conclusion is generally appropriate, however, only for local effects which often occur soon after exposure has occurred.

4.1 ABSORPTION AND DISTRIBUTION

Why is this the case? Once the chemical enters the body through one of these routes it has the potential to be absorbed into the bloodstream. If this occurs, the chemical is then transported throughout the body and all of the organs and tissues have the potential to be exposed. If one or more of these parts of the body is more sensitive than the site of entry, more severe toxic effects may occur to that organ or tissue. If the material remains in the blood for a long period of time, there will be repeated exposure of the tissues and organs through

23

which the blood passes. Thus a single external exposure can lead to repeated internal exposures and possible toxicity to a number of organs or tissues.

Once a substance is absorbed and gets into the bloodstream, it has a variety of initial fates. It may be excreted as is; it may be stored as is; or it may interact with the body chemicals and be altered in some way. This last process is called metabolism, and the altered products that are formed as a result of metabolism are called metabolites. A chemical that is not altered by the body will not have metabolites; one which interacts with the body chemicals will have one or more metabolites. Generally, metabolism is not 100% efficient, so that some of the original chemical will remain unchanged. Each of the metabolites has the same variety of fates as the original absorbed material. Thus exposure to one chemical may result in the excretion or storage of several different chemicals as well as the potential for a variety of toxic effects. It should be emphasized that these are usually not either/or possibilities—absorbed substances and their metabolites are often partially excreted, partially stored, and partially available to produce adverse effects.

4.2 METABOLISM

Metabolism is important in a number of body processes, one of which is the detoxification of foreign chemicals. One main metabolic detoxification mechanism is the conversion of undesirable chemicals, which are often hard to excrete, into less toxic chemicals which can be readily excreted by the body. The kidney, the main excretory organ of the body for foreign chemicals, is most efficient at eliminating polar molecules. It does not function well with very nonpolar compounds, and so these tend to remain in the body for long periods of time. Thus metabolic processes that are detoxifying in nature generally involve reactions that convert nonpolar molecules into more polar ones. In most cases, these changes are advantageous for the body, but in some instances this process converts basically nontoxic nonpolar compounds into more toxic polar (or more polar) metabolites. An example of a metabolic process which illustrates these points is given in Figure 4.1.

The chemical alterations that are the basis of metabolism are due to the action of enzymes. Enzymes are proteins with very particular chemical and physical properties which allow them to recognize and attach to particular types of molecules. They then facilitate the interactions that are responsible for the metabolic changes. Specific enzymes that recognize particular types of molecules are normally present in small quantities, and the body manufactures more of them when the need arises, e.g., after a significant exposure to the appropriate foreign molecule. This process is not easily reversible, and the body does not revert back to its preexposure status very rapidly. Thus exposure to one chemical of a particular type may lead to the presence of a large number of specific enzymes when a subsequent exposure to the same or a similar chemical occurs. This process of increasing the enzyme levels is called enzyme induction. In general, this is beneficial since it helps the body to respond rapidly to repeated exposures. Of course, if the metabolic processes resulting from the response lead to more toxic rather than less toxic metabolites, this induction is counterproductive.

Non-Polar   **Toxic Intermediate**   **Polar**

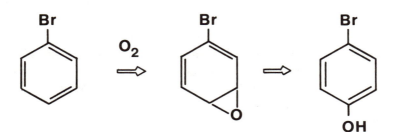

Figure 4.1   Example of metabolic changes leading to more easily excretable compound.

4.3   STORAGE AND EXCRETION

After metabolic processes have occurred, mainly in the liver and the kidney, both the original compounds to which the body was exposed, and possibly one or more metabolites, circulate through the bloodstream. As indicated previously, they may be stored or excreted, as well as produce toxicity. Storage may occur in the blood or in other parts of the body. The place that chemicals are stored and the percent that meets this fate depends mainly on the characteristics of the chemical. Nonpolar compounds, such as the organochlorines (DDT, PCBs, chlordane, etc.), are stored mainly in the fat. Other chemicals, which are recognized and bound by blood proteins, especially albumin, can be stored in the blood. Still others, such as lead, may be stored predominantly in bone.

Although the kidney is the main excretory organ, chemicals or their metabolites may also be eliminated from the body in other ways. The second most common route is through the feces. Other generally less important routes are through the lung and through sweat or saliva. In females, there are two additional alternatives. Chemicals may be excreted through breast milk and through the eggs. In these situations, excretion from the mother's viewpoint represents exposure from the offspring's perspective.

A last fate, in addition to storage and excretion, is for the compound to react with the body to cause some beneficial or deleterious effect. In the case of drugs, the major effect is beneficial, but all medication also has the potential for side effects, i.e., deleterious interactions with the body. Indeed, an important impetus for toxicology research stemmed from this recognition of the adverse effects that largely beneficial chemicals can have on individuals. At present, however, many toxicologists are greatly concerned about substances that have no beneficial health effects, only adverse ones. The following discussion focuses on this type of compound.

4.4 RECEPTORS

The question of exactly how environmental contaminants cause toxicity cannot be answered in great detail. The symptoms have been clearly delineated in many cases, and the organs that are affected have been determined, but the sequence of events which leads from exposure to adverse effects is not well understood. This is particularly true for chronic or long-term effects, such as cancer. The present understanding is that the first step in many, but not all, cases involves recognition of the toxic chemical by a specific molecule or group of molecules in the body, known as the receptor. This recognition is probably attributable to both the chemical and physical properties of the toxicant. It is thought that the receptor must have a shape which is complementary to that of the toxicant so that a close fit occurs. The usual analogy is to a lock and a key. It should be noted that the idea that receptors exist has not been verified by the ultimate test, the isolation and identification of a receptor molecule. Figure 4.2 shows a group of chemicals that have similar structure and cause similar toxic effects, and the general shape of the postulated receptor for these chemicals. The exactness of fit is related to the toxic potency, with TCDD exhibiting the greatest toxic potential.

The binding of the toxicant to the receptor is thought to initiate a chain of events that leads to the ultimate adverse effect. Thus the location of these receptors has a strong influence on the site of toxicity. The number of receptors influences the degree of toxicity. The steps that follow the initial toxicant-receptor interaction are not known. It appears that binding to the receptor is reversible, since removal of the source of exposure will often lower the blood concentration of the toxicant and in turn lead to release of some toxicant molecules from attachment to the receptor and back into the bloodstream. Over time, a balance between bound and unbound chemical will be established. This balance will, of course, change if new

TOXIC CHEMICALS

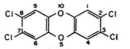

2,3,7,8-Tetrachlorodibenzo-p-dioxin

(TCDD)

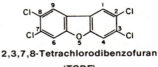

2,3,7,8-Tetrachlorodibenzofuran

(TCDF)

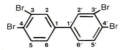

3,3′,4,4′-Tetrabromobiphenyl

(PBB)

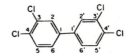

3,3′,4,4′-Tetrachlorobiphenyl

(PCB)

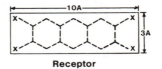

Receptor

Figure 4.2 Polycyclic aromatic compounds that produce similar types of toxic effects and a diagram of the postulated receptor site.

exposures occur. A schematic representation of the sequence of events following exposure is given in Figure 4.3 for two different toxicants.

The events that lead up to toxicity can be summarized as exposure, absorption, and distribution of the toxicant; metabolism of the toxicant; and beneficial or adverse interaction of the toxicant, and any metabolites, with the body. Of the amount absorbed, the fraction that is excreted is clearly not available to cause toxicity. The part that is stored is also unavailable to produce adverse effects. It is the last fraction, which is usually bound to the receptor, that is responsible for toxicity. Unless all of the receptors are used up, i.e., bound to toxicant, the addition of more toxicant will lead to a greater

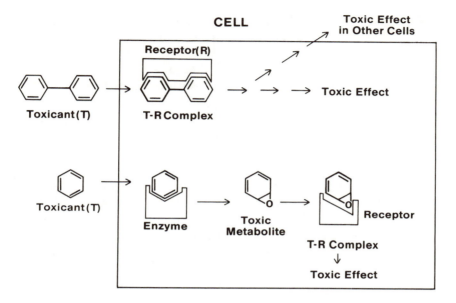

Figure 4.3 Possible pathways between cell exposure and expression of toxicity.

number of toxicant-receptor complexes and thus greater toxicity.

4.5   TIME COURSE OF TOXICITY

In addition, if the toxicant that is initially stored is some-how remobilized and circulates again through the body, reex-posure may occur. This may happen, for example, if an individual with a toxicant stored in the fat loses weight rapidly. In such a case, it is possible to raise the blood concentration of the toxicant as the number of available storage sites decreases and toxicant is released. Additional toxicant-receptor interactions at sites of toxicity could then occur as a new balance is established.

The above discussion illustrates that toxicity can vary over time even in the simple case where there is only a single exposure. The situation with respect to multiple exposures to the same chemical is more complex. In the simplest case, repeated exposures may have no more effect than a single exposure. This would be the case if each dose is too low to cause any adverse effects and each dose is excreted by the body before another dose is administered. Thus a worker, for example, might be exposed to the same dose daily and show no effects if this dose was eliminated by the body between the end of one work day and the beginning of the next.

A recent situation illustrates the importance of the condition stated in the previous paragraph. In this case, the work schedules were altered and the employees worked 12-hour days instead of 8-hour days. Toxic effects started to appear in the workers soon after this schedule was adopted. This was not due to the longer exposure time but to the decreased time between work periods. Excretion was not complete during this period, so that the chemical concentration in the blood increased from day to day, finally reaching a toxic dose. This example illustrates the critical importance of the timing of exposure to the possibility that adverse effects will occur. It is not only the total exposure that counts but also the frequency of exposure. This concept is illustrated graphically in Figure 4.4.

In summary, the dose-response that was discussed earlier can be traced in many cases to the formation of toxicant-receptor complexes. A higher dose leads to more complexes and thus greater toxicity. The toxic effects are due to a series of chemical interactions that are not yet understood and involve molecules that have not yet been isolated. Gradually toxicologists will be able to fill in the links in the chain of events and describe exactly what happens from the instant of exposure to the time of appearance of adverse effects. In the meantime, it is fruitful to utilize our current understanding to analyze the types of adverse effects that are known to occur. This is the subject of the next chapter.

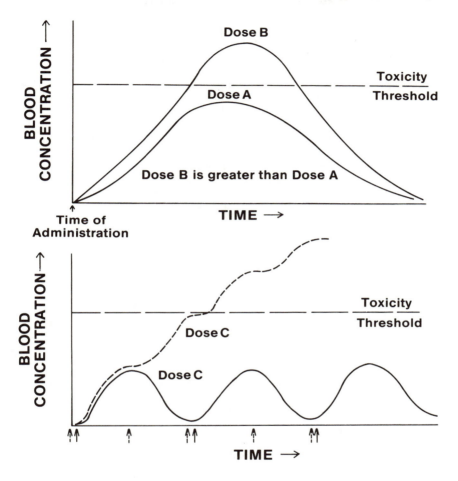

Times of administration indicated by ↑ for solid curve and ↑ for dashed curve. The same dose is administered at each time.

Figure 4.4 Relationship between time course of exposure and toxic expression.

5

Types of Toxic Effects

Although toxic effects are divided into acute, subacute, and chronic on the basis of time course, it is not possible to categorize each type of toxicity or the effects of individual chemicals into one of these three classifications. For example, lung toxicity may be an acute, subacute, or chronic effect. Similarly, a particular chemical may produce an acute effect at one exposure level and a subacute or chronic effect at another.

Despite these classification difficulties, it has been common practice to identify the different types of toxicity with one of the three time-course categories. Thus organ damage has usually been classified as an acute or subacute effect, and most other types of toxicity have been assigned to the chronic classification. The effects generally included in the chronic category are carcinogenesis, teratogenesis, reproductive toxicity, and mutagenesis.

5.1 ORGAN TOXICITY

Organ toxicity covers a very wide range of organ systems and toxic effects. In addition to the heart, lungs, kidneys, and liver, toxic effects can also decrease functioning of the pancreas, spleen, thymus, and even nondiscrete organs, such as the skin. In many cases, the effects on one organ system may be manifested throughout the body. For example, depression of thymus function may reduce the efficacy of the immune system and so make the organism susceptible to infectious diseases. Effects may be less general in other cases such as skin lesions, e.g., chloracne.

Organ toxicity has generally been thought of as an acute or subacute effect for a number of reasons. One is that the effect often occurs soon after the initiation of exposure, especially at high doses. In addition, the degree of toxicity reflects the frequency and intensity of exposure in the traditional dose-response fashion described earlier, so that quantitative measures of toxicity can readily be determined. A final characteristic of organ toxicity is that the organ that is affected is directly linked to the type of chemical to which exposure has occurred.

5.1.1 Lung Toxicity

For example, the lung can be acutely affected in a variety of ways by toxic substances. Inhalation of high concentrations of oxidizing gases, such as oxygen, ozone, or nitrogen dioxide, can cause rapid direct damage to lung cells and subsequent loss of lung function. The lung can also be affected indirectly by chemicals that are ingested or pass through the skin and reach the lung through the circulation. Paraquat, for example, can reach the lung after absorption through the skin or gastrointestinal tract and cause lung cell injury.

While the above examples are consistent with the identification of organ damage with acute toxicity, the effects of other exposures are not. For example, one lung disease that results from smoking, emphysema, is the result of long-term exposure to the toxicant. Similarly, another lung disease, asbestosis, results only after prolonged exposure to asbestos. Thus these organ toxicities are chronic rather than acute effects. In addition, it does not appear that these effects follow the simple dose-response relationship discussed earlier. This is not surprising, since it is presently thought that asbestos, at least, exerts its toxic effects more through its physical than its chemical properties. Such a mechanism of action would be unlikely to involve a receptor and thus would not be describable in the same way as the action of traditional toxicants.

5.1.2 Kidney Toxicity

Another organ which is commonly affected by toxicants is the kidney. Just as the lung is vital for one bodily function, respiration, the kidney is vital for another, the excretion of undesirable chemicals. Toxicants which interfere with kidney function can produce severe adverse effects. One well-studied toxicant that affects the kidney is lead. It can inhibit functioning of the parts of the kidney where toxicants are filtered out from the blood and so can lead to the retention of undesirable chemicals and subsequent increases in concentrations of these chemicals in tissues and organs throughout the body. Eventually toxic levels may be reached and adverse effects will ensue. In this case, a variety of organs may be affected, not only the original site of toxicity.

5.1.3 Multiple Effects

Lead is a toxicant that illustrates another point: Not only can it indirectly affect a number of organs, it is directly toxic to more than one body system. It is toxic to the nervous system as well as the kidneys. Lead causes changes in nerve cells that affect their ability to transmit nerve impulses properly. As a result, nervous control over muscle action may be decreased and palsy or shaking can occur. The effects on the nervous system are cumulative, so the symptoms may not appear until exposure has occurred over a long period of time. Thus lead is another organ toxicant that can cause chronic toxicity.

Lead also serves to illustrate the differences in toxic effect that are manifest when exposure occurs at different ages. Children have been shown to be especially sensitive to the effects of lead, and there is evidence that low lead exposure at an early age can lead to permanent mental deficits in the exposed individuals. It is thought that the gradual development of the brain in the years after birth makes this organ particularly sensitive to toxicants which can interfere with developmental processes. This type of toxic effect is largely responsible for the great concern about lead in gasoline and drinking water and the steps that have been taken to limit the levels of this chemical in fuel.

The above examples of organ toxicity reveal the complexity of the subject and the difficulty in trying to classify toxic effects into precise categories. Some organ toxicities occur only after long periods of exposure, so the identification of organ toxicity as an acute or subacute effect must be heavily qualified. In addition, it can be seen that a single chemical may cause multiple organ effects, some of which depend on the age of the exposed individual. Thus toxicity to a specific organ may not be uniquely associated with a particular time course or a specific chemical.

5.2   CARCINOGENESIS

Cancer is the chronic toxic effect that is of most concern to the general population and thus the most well studied. As indicated earlier, it is considered a chronic effect, although it may appear after shorter exposures that fall into the subacute category. A more important distinction is that the severity of the effect (cancer) is not related to exposure in a dose-dependent manner, although the frequency of effect is. That is, the proportion of individuals who get cancer is related to the degree of exposure, but the extent of the tumor spread or growth is independent of exposure.

Another important difference between cancer and organ toxicity is that carcinogenesis results from abnormalities in the reproduction and growth of cells rather than alterations in cell structure or function. A tumor results from uncontrollable cell proliferation and may result from toxic alterations of only a single cell. Thus another distinction from organ toxicity is that carcinogenesis may not involve a threshold, or minimum dose.

One characteristic that may be shared in many cases between organ toxicity and carcinogenesis is that specific chemicals may cause specific types of toxicity. Thus a substance discussed previously in this chapter, asbestos, leads to lung cancer as well as asbestosis. A third toxic effect, the occurrence of a unique and rare form of cancer, mesothelioma (a cancer of the lining of the lung), has also been linked to asbestos since it occurs almost exclusively in asbestos-exposed individuals. It does, however, occur at a much lower incidence than lung cancer. Another example of a specific chemical causing a specific effect is vinyl chloride. The link between this compound and cancer was revealed through the occurrence of a very rare type of liver cancer, angiosarcoma, in workers who were heavily exposed to this chemical. These examples represent unique situations where human cancer can be linked to a specific chemical. The limited number of generally accepted human carcinogens are listed in Table 5.1.

Overall, while organ toxicity and carcinogenesis share some similarities, the differences clearly place them in distinct categories. As will be shown later in this chapter, these differences reflect very distinct types of toxicological events.

Table 5.1   Human Carcinogens

4-Aminobiphenyl
Analgesic mixtures containing phenacetin
Arsenic and certain arsenic compounds
Asbestos
Azathioprine
Benzene
Benzidine
N,N-bis(2-chloroethyl)-2-naphthylamine (chlornaphazine)
Bis(chloromethyl)ether and technical grade chloromethyl methyl ether
1,4-Butanediol dimethylsulfonate (myleran)
Certain combined chemotherapy for lymphomas
Chlorambucil
Chromium and certain chromium compounds
Coke oven emissions
Conjugated estrogens
Cyclophosphamide
Diethylstilbestrol
Hematite underground mining
Isopropyl alcohol manufacturing (strong acid process)
Manufacture of auramine

Table 5.1, continued

Melphalan
Methoxsalen with ultraviolet A therapy (PUVA)
Mustard gas
2-Naphthylamine
Nickel refining
Rubber industry (certain occupations)
Soots, tars, and mineral oils
Thorium dioxide
Vinyl chloride

From: 4th Annual Report on Carcinogens, National Toxicology Program, U.S. Department of Health and Human Services.

5.3   TERATOGENESIS

Teratogenesis is the formation of birth defects in offspring, often as a result of maternal or paternal exposure to a toxicant. These abnormalities can arise in a number of ways but most commonly result from alteration of the developing cells, leading to improper functioning of these cells or interference with differentiation so that the proper cell types do not form or do not form in the proper number or location. It is probable that the types of events that result in teratogenesis also may result in death of the embryo or fetus. Thus there will be no birth defect, as there will be no birth. This provides another example of the difficulty in precisely defining toxic effects in a single category. As will be seen in the next section, such fetal or embryonic death generally falls into the category of reproductive toxicity, and indeed, teratogenesis is sometimes included as one form of reproductive toxicity.

Teratogenesis is usually classified as a chronic effect, although the toxicity appears within a relatively short time as compared to the lifetime of the individual. Indeed, it might more properly be labeled a subacute effect on the basis of time course. However, some of the birth defects that result from toxicant exposures are heritable and may appear in future generations as well as the present one. In this case, the time course is much longer, but, since the exposure is brief, the effects might be more properly described as delayed rather than chronic.

The classification of teratogenesis as a chronic effect is based more on the number of important features it shares with carcinogenesis. For example, the defects may be the result of a single exposure. In addition, the severity of the effect may be unrelated to the degree of exposure, although the frequency of effect does show a simple dose dependence. Teratogenic agents often are specific so that a particular type of birth defect is related to a specific chemical exposure. One well-known case, exposure of pregnant women to the drug thalidomide, resulted in a well-characterized group of severe birth abnormalities. It is probable, however, that not all toxicants will lead to such specific teratogenic effects, and some will cause a variety of toxic events.

Thus teratogenesis shares a number of characteristics with carcinogenesis and has few similarities to organ toxicity. As a result, classification in the same general category as carcinogenesis is plausible. However, it should be recognized that the time course of teratogenesis is inconsistent with this categorization.

5.4 REPRODUCTIVE TOXICITY

Reproductive toxicity is a broad category that includes a variety of effects on the reproductive capacity of living systems. These effects can involve decreases in fertility, decreases in percent of conceptions leading to live birth, or fetal or embryonic toxicity. This last category may be distinguished from teratogenicity in that the toxicity does not lead to birth defects, but instead, may lead to reduced birth weight or size. Two common toxicants that can lead to such fetotoxicity are alcohol and tobacco. As indicated in the previous section, teratogenic agents may also lead to fetal death in some situations, so an agent that is a teratogen may also cause reproductive toxicity.

The variety of effects that are included in reproductive toxicity makes time-course classification difficult. Agents which affect the fertility of either males or females may cause their effects only after long periods of exposure. On the other hand, some reproductive toxicants may produce their effects after short-time exposure of the mother during gestation. The effects, however, may not be manifest until a later time. Thus some types of reproductive toxicity can be considered chronic, other types subacute, and still others delayed on the basis of time course.

Some of the subacute reproductive effects are clearly more similar in nature to carcinogenesis and teratogenesis than to organ toxicity. They show limited relationship between dose and severity of toxicity and the absence of a distinct threshold of toxicity. Thus it is not too surprising that reproductive toxicity is classified together with carcinogenesis and teratogenesis in the category of chronic toxicity.

5.5 MUTAGENESIS

Mutagenesis is the formation of mutations or changes in the genetic material. These changes may affect the individual in which mutations occur or they may affect later generations if the mutations occur in the genetic material which is passed on to offspring. The effects that occur in the individual exposed to a mutagen are often long-term, but high doses may cause toxicity, which occurs in the subacute time range. Obviously, the effects which occur to future generations are long-term, but it would probably be more accurate to call them delayed rather than chronic effects, since they may occur after a single high exposure rather than multiple exposures over a long period of time.

Mutagenesis shows a number of characteristics in common with carcinogenicity and some with teratogenesis and reproductive toxicity. For example, the frequency of the toxic effect is related to extent of exposure but not necessarily the severity; i.e., the extent of the toxic effect may be related to the type of alteration which has occurred rather than the degree of exposure. Another similarity to carcinogenicity is that mutagenicity may cause its effects through the reproduction or proliferation of abnormal cells rather than loss of function, as is common in organ toxicity. A third similarity is the possible absence of a threshold. It is conceivable that a single mutational event could lead to overt toxicity.

The similarities between mutagenesis and the other types of toxicity that are classified as chronic is not coincidental. Mutagenesis is thought to be the primary event leading, or at least contributing, to the other types of effects in at least some cases. In the case of carcinogenesis, the link is thought to be quite strong for many chemicals. In these situations, mutagenesis is believed to alter cells in such a way that they have the potential to grow into malignancies. This initial step is called initiation, and the mutagenic agents that cause this step are called initiators. Initiated cells do not always become cancers, indicating that other events are generally necessary for these genetically altered cells to start proliferating abnormally. This second step is called promotion, and some toxicants appear to act only as promoters and thus lead to cancer only when genetically altered cells are already present. It is thought that multiple and chronic exposures to promoters are needed to produce tumors. The two-step theory of carcinogenesis is illustrated in Figure 5.1.

It should be emphasized that carcinogenesis can occur by a variety of mechanisms. The way that asbestos, a physical agent, leads to cancer is undoubtedly different from the mode

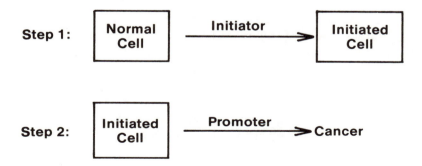

Figure 5.1 Two-stage model of carcinogenesis.

of action of a chemical carcinogen, such as vinyl chloride. In addition, there is clear evidence that not all chemical carcinogens work by the same mechanism.

Teratogenesis can also result from a mutagenic event. In this situation, the genetic material of a cell is altered so that when it divides and multiplies, it leads to an organ or tissue which cannot function properly. It is also possible that a mutagenic event will affect molecules that are necessary for normal differentiation and so may interfere with the correct development of the embryo or fetus. Mutagenesis, however, is not required for teratogenesis. A toxicant can block normal development by direct toxicity to developing cells or by interfering with the sequence of chemical events necessary for the correct number and placement of cells in the offspring. Thus some types of teratogenic effects will share characteristics with mutagenesis and carcinogenesis, and others will not.

Reproductive toxicity can also result from a mutagenic event, although this is not too common. An example of a situation where this might occur is if the mutagenic event is such that it leads to death or resorption of the embryo or fetus. It is also possible that mutagenic events could lead to abnormalities of the egg that could affect fertility. With the limited number of eggs in the female, there is the possibility of a limited number of mutagenic events having a measurable impact on fertility. In any event, most reproductive toxicity is probably due to more direct effects of toxicants and would not share the characteristics of mutagenesis.

Thus it can be seen that the events that can follow a mutation are many. Since the mutational effect is the initial one, the characteristics of the overall toxicity will reflect this initial event. Thus it is not surprising that carcinogenicity, teratogenicity, and reproductive toxicity share a number of characteristics with mutagenicity, at least in some cases. The most important of these are the lack of a relationship between intensity of effect and degree of exposure; the potential lack of a threshold; and the importance of growth and proliferation of abnormal cells to the final effect.

Overall, it can be seen that there is no way to simply group toxic effects that may have similar symptomology. Since the mechanism behind the similar-appearing effects may be different, the toxicity must be regarded as distinctive and must be understood in this context. For convenience, it has been the practice to classify toxic effects in broad categories related to their time courses. This has been useful in some ways, especially with respect to testing, but as we have learned more, it has become clear that the categories may conceal more than they reveal in some cases.

6

Toxicity Assessment

Recognizing all of the variables that can affect the toxicity of a particular chemical, and also the tremendous variety of organs and tissues that can be affected, it is not surprising that assessing the toxicity of specific chemicals is a complex and uncertain undertaking. This is especially true when it is the toxicity to humans that is under investigation. The obvious impossibility of controlled experiments on humans means that the uncertainties mentioned above are exacerbated by the need to use surrogates, especially rodents.

Toxicity testing is currently performed in a number of different ways, many of which grew historically in response to particular concerns and regulations. Tests for acute, or short-term, toxicity were developed largely in response to concerns about the most highly exposed groups, mainly workers. On the other hand, testing for long-term, chronic effects has evolved in response to concerns about the general population as well as subgroups. The most well-developed chronic test, the carcinogenesis bioassay, has gradually evolved from a test to determine if a chemical is or is not a carcinogen to one aimed at quantitatively establishing the dose at which the chemical is expected to produce a specific cancer incidence in an exposed human population.

6.1 ACUTE TOXICITY TESTING/LD$_{50}$

Perhaps the simplest and most commonly applied toxicity test is the single-exposure study, with death as the criterion of toxicity. This study is performed by administering a single dose to a limited number of animals and determining the number of deaths after 14 days. It is designed to determine the LD_{50}, or lethal dose for 50% of the animals. This test explicitly reveals that even in carefully bred laboratory animals, there is so much variation that a particular dose will kill only a certain percentage of the population. In addition, LD_{50} studies performed with the same chemical but exposing populations of other types of animals reveal wide variations among species. The LD_{50} for one type of animal could be thousands of times less than that for another type of animal. In other words, a very small amount might kill 50% of the guinea pigs studied but none of the rats. It might require thousands of times more chemical to kill 50% of the rats. Table 6.1 shows the ranking of LD_{50}s for a number of "common" and "toxic" chemicals.

In order to utilize LD_{50} studies to provide a meaningful comparative ranking of chemicals, it has been the practice to study one standard species, the rat. While the LD_{50} does provide a measure of the amount needed to kill rodents, the values

Table 6.1   Approximate Acute LD_{50} of Selected Chemical Agents

Agent	LD_{50} (mg/kg)
Sugar	29,700
Polybrominated biphenyls (PBBs)	21,500
Alcohol	14,000
Methoxychlor	5,000
Vinegar	3,310
Salt	3,000
Malathion	1,200
Aspirin	1,000
Lindane (Benzene Hexachloride delta isomer)	1,000
2,4-D	375
Ammonia	350
DDT	100
Heptachlor	90
Arsenic	48
Dieldrin	40
Strychnine	2
Nicotine	1
Dioxin (TCDD)	0.001
Botulinus toxin	0.00001

cannot be extrapolated to provide an estimate of how much of the material would kill 50% of the humans exposed to it. However, it is useful in predicting relative toxicity in humans and thus for assigning chemicals to broad classes for purposes of regulation. Pesticide labels, for example, contain signal words which reflect the relative LD_{50}s of the active ingredients. The words *Danger-Poison* designate the most toxic class; *Warning* indicates the pesticides of intermediate toxicity; and *Caution* those of lowest relative toxicity. Table 6.2 shows how these categories are defined.

In addition to providing a relative ranking, the studies that are performed to determine the LD_{50} can provide clues about toxic effects that might occur in longer-term tests. If careful observations of the animals are made during the course of

Table 6.2 Toxicity Categories for Pesticides

	Toxicity Categories (Signal Words)			
	I (Danger-Poison)	**II** (Warning)	**III** (Caution)	**IV** (Caution)
Oral LD_{50}	Up to and including 50 mg/kg	From 50 thru 500 mg/kg	From 500 thru 5000 mg/kg	Greater than 5000 mg/kg
Inhalation LD_{50}	Up to and including 0.2 mg/liter	From 0.2 thru 2 mg/liter	From 2 thru 20 mg/liter	Greater than 20 mg/liter
Dermal LD_{50}	Up to and including 200 mg/kg	From 200 thru 2000	From 2000 thru 20,000	Greater than 20,000
Eye effects	Corrosive; corneal opacity not reversible within 7 days	Corneal opacity reversible within 7 days; irritation persisting for 7 days	No corneal opacity; irritation reversible within 7 days	No irritation
Skin effects	Corrosive	Severe irritation at 72 hours	Moderate irritation at 72 hours	Mild or slight irritation at 72 hours

these experiments, effects that precede death in the high-dose animals may be noted. It may also be possible to detect adverse, but not lethal, effects in the animals receiving lower doses. Some clinical indications of acute toxicity are listed in Table 6.3.

Table 6.3   Some Clinical Indications of Acute Toxicity

Structure or Function Affected	Possible Effects
Respiratory	Changes in rate or depth of breathing
Motor activities	Changes in frequency and nature of movement
Reflex	Response to external stimuli
Ocular	Tearing
Cardiovascular	Changes in heart rate
Gastrointestinal	Vomiting
Dermal	Swelling or redness

6.2   SUBACUTE TOXICITY TESTING

The LD_{50} is a very crude measure of toxicity because it does not recognize any toxic effect short of death. Obviously, death is not the only type of toxicity that needs to be avoided, and thus other tests are employed to examine toxic effects that are less extreme. The most common of these tests is a subacute (or subchronic) assay that assesses the effects in an animal population of daily exposure to a toxicant over about 10% of the animals' lifetime. In rats, this corresponds to about three months. Careful study of the animals, including examination of all body tissues and fluids, reveals the dose at which toxic effects begin to occur. Table 6.4 is a list of the studies that are part of the protocol for subacute studies.

In distinction to the LD_{50}, where only one endpoint, death, is used, the subacute study looks for many possible endpoints. These might be organ dysfunction, behavioral changes, or alterations in levels of normal body fluid components. Also distinct from the LD_{50}, the highest dose at which none of the animals shows toxic effects is determined. This **no observable adverse effect level** (NOAEL) provides a quantitative meas-

Table 6.4   Indications of Subacute or Chronic Toxicity

Hematology and clinical chemistry
Urinalysis
Organ weights: Liver, kidneys, heart, gonads, brain, adrenals
Histopathology: All tissues*

*All tissues include:

Gross lesions	Small intestine (duodenum,
Skin	jejunum, ileum)
Mandibular and mesenteric lymph	Large intestine (cecum, colon,
node (all studies)	rectum)
Bronchial and mediastinal lymph	Tissue masses or suspect tumors
nodes for inhalation studies	and regional lymph nodes
Mammary glands with adjacent skin	Liver
Salivary glands	Gall bladder (mice)
Thigh muscle	Pancreas
Sciatic nerve	Spleen
Sternebrae, femur, or vertebrae	Kidneys
including marrow	Adrenals
Costochondral junction, rib	Urinary bladder
Thymus	Seminal vesicles
Oral cavity, larynx and pharynx	Prostate
Trachea	Testes, epididymis vaginal tunics
Lungs and bronchi	of the testes and scrotal sac
Heart and aorta	Ovaries
Thyroid	Uterus
Parathyroids	Pituitary
Esophagus	Spinal cord
Stomach	Eyes
Tongue	Preputial or clitoral glands
	Zymbal glands (auditory
	sebaceous glands)

ure of the toxicity of each chemical in each animal species. Again, there is variability among species so that one type of animal may show toxic effects at dosages much lower than another type.

This type of study is designed to provide not only a relative ranking of chemicals but also a quantitative estimate of the levels at which toxicity will occur in humans who are exposed. Since the experiments are performed on animals, some factor must be included to take into account the possible differences in response between humans and other species. Actually, because the object of the testing is to determine a level in humans which provides some margin of safety, a number of features that are not part of LD_{50} studies are part of the testing protocol. Instead of using one species, rats, as the model to extrapolate from, several species are studied, and the most sensitive one is chosen as the human surrogate. Then, the highest dose at which this most sensitive species shows no effect is generally divided by 100 to account for possible greater sensitivity in humans than in the test species and for greater variation in human populations, as opposed to inbred laboratory animal populations.

It should be emphasized that the level that is calculated by this procedure is not the best estimate of the NOAEL in humans but instead a value designed to ensure some margin of safety. This final number is often called the safe level, and the factor of 100 the safety factor. However, safe is not defined in any quantitative manner. It is often assumed by members of the public that it means absolute safety, i.e., no adverse effect possible in any individual. This is not the case, since the studies on which this level are based are done on a limited number of animals, and it is quite possible that an effect would be noted in one or more animals if a larger population was studied under the same exposure conditions. What this level really means is difficult to articulate in simple terms. A generally accepted interpretation is that there is a small likelihood of adverse effects if humans are exposed at this level and also that this likelihood is not significantly increased by slightly exceeding this level. It is certainly not the case that this safe value represents the demarcation line between no effect and some effect in humans.

It should also be noted that the NOAEL does not provide any indication of the seriousness of the toxicity that occurs above the no-effect level. The effects may be local in nature, e.g., skin irritation, or may affect the whole organism, e.g., central nervous system damage. The effects may be reversible or irreversible. Thus the NOAEL provides a quantitative measure of the dose at which an effect may occur but does not provide even a qualitative indication of the clinical significance of this effect.

6.3   CHRONIC TOXICITY TESTING/ ANIMAL EXPERIMENTATION

Chronic toxicity testing is done in a number of ways, but the study that is most commonly done and most closely examined is the rodent carcinogenesis bioassay. In this type of study, groups of rats or mice are exposed to a particular chemical every day (in most cases) at the same dose for a lifetime. During the course of and at the end of the study, the animals are examined to check for tumors, and the incidence at each dose is recorded. Generally, two or three doses are used to establish a dose-response relationship for cancer.

There are a number of differences between this assay and the acute and subacute tests. One important one is that the studies are performed with very high doses compared to the NOAEL, but lower doses than in the LD_{50}. The highest dose used is called the maximum tolerated dose (MTD) and can be defined as the level of exposure that will not shorten the lifetime of the animal due to toxic effects other than cancer. The other doses that are studied are also very high. The reason for this type of exposure is that cancer is expected to occur in a very small percentage of the population at doses that might be found in the human environment. Thus a very large number of animals, i.e., millions, would have to be studied to find cancer in a significant number of the animal population at such realistic doses. This is not feasible, so that high doses are employed to produce cancer in a high percentage of reasonably sized groups (about 50 animals), e.g., 30% to 70% cancer incidence.

In order to determine the cancer incidence at the levels of concern for humans, the high-dose data must be mathematically manipulated to calculate a low-dose value. Unfortunately, there is no scientific consensus about the mechanism of cancer production and thus no general agreement as to the mathematical formula that should be used to make the extrapolation. Depending on the assumptions, a variety of curves could be drawn, and a tremendous range of values could be determined. The calculated dosage corresponding to a particular cancer incidence could vary by several orders of magnitude depending on the assumptions which are employed. Figure 6.1 shows curves generated with a number of different models.

Since the purpose of the testing is to determine maximum permissible levels of human exposure, very conservative assumptions are used to estimate the minimum doses or exposures that will produce cancer. That is, the data from the high dose studies are extrapolated in such a way as to estimate the highest possible cancer incidence for a given dosage. This again follows the principle discussed with the NOAEL measurements, that of providing some margin of safety. In general, the bioassay is used to determine the dosage that will produce one cancer in 100,000 or 1,000,000 humans if daily exposure to the same dose occurs over a lifetime. Since this value is generally based on the assumption that leads to the lowest dosage, it does not represent the best value. Thus the dosage that is most often quoted is not some exact level, but

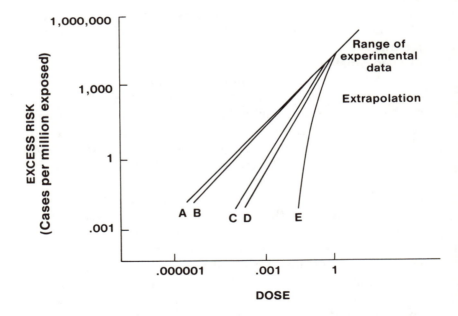

Figure 6.1 Low-dose extrapolation for a carcinogen under several mathematical models: A, B, C, D, and E.

just a conservative value that is expected to provide some margin for error.

The use of this type of bioassay is more controversial than the NOAEL not only for the reasons just stated, i.e., the uncertainties in extrapolation. In addition, there is serious question as to the relevance of high-dose effects to those which occur at lower doses. It is possible, for example, that high doses could overwhelm the organism's response system while low doses would not. Thus, the quality of the response would be different, not only the quantity. There are other controversies that have yet to be resolved also. For example, it appears that there are at least two different types of carcinogens, initiators and promoters, which act by different mechanisms. Can the same experiments and the same extrapolations possibly provide valid results for both? In response to such concerns, there are many suggestions for changes in the way this type of testing is performed. However, adoption of new approaches will be slow in coming if history is any guide.

6.4 CHRONIC TOXICITY TESTING/EPIDEMIOLOGY

A fourth important type of toxicity testing method is epidemiology. This is the study of human populations to establish correlations between particular environmental conditions and specific health effects. In the type of situation most commonly of interest, epidemiological studies are used to establish links between exposures to particular toxic substances and specific chronic effects. A good example of this type of study is the epidemiological investigations of smokers that suggested a link between this habit and lung cancer and other diseases. In this case, it was possible to show that the incidence of the toxic effects was related to extent of smoking. It took a long time to reach this conclusion even though this was a fairly clear-cut case. Other examples of successful epidemiologic studies are those linking asbestos with a variety of chronic ailments, vinyl chloride with a rare form of liver cancer, and benzene with leukemia.

The basic problem with epidemiological studies is that they are uncontrolled and are done on a very variable population. They are uncontrolled in that there are a tremendous variety of exposures that people have, and it is impossible to control any one of them to determine what effect that change will have. Thus it is very rare that the one important factor can be picked out from all of the others. Population variability is very important also. If an effect occurs in a very small percentage of the population, it will be difficult to pick out from the tremendous range of effects that would be shown even in the absence of the suspect chemical.

6.4.1   Retrospective/Case-Control Studies

There are basically two types of epidemiological studies. The first type is the case-control, or retrospective type, where two populations are studied. The first population consists of individuals who demonstrate the toxic effect of interest, and the second is made up of those who do not. The two populations are matched as well as possible with respect to all other variables, e.g., age, socioeconomic status, and so on. Then the past histories of exposure of the two populations are investigated to see if some differences can be determined which might be related to the toxic effects observed. This type of study depends heavily on information which is hard to gain. One important type of evidence comes from the recollections of the individuals under study and these may well be selective or otherwise unreliable. Another possible source of evidence, medical or occupational records, is more tangible but also generally incomplete. Thus, for example, even if some records were kept in the workplace, many factors which may be important were not measured. In addition, records from one workplace may not be comparable to those from another, adding to the difficulty of combining the evidence in any meaningful way. It is not too surprising, therefore, that such retrospective studies have rarely provided even a qualitative estimate of the chronic toxicity of a particular chemical.

6.4.2 Prospective/Cohort Studies

The second type of epidemiological study, the prospective or cohort study, follows a population from a set time into the future. Although this eliminates the need for recall, it can introduce other problems. One is that individuals who are taking part in the study may act in ways that make interpretation difficult. For example, they may be more conscious of their health and develop eating and exercise patterns quite different from the norm. Another problem is that it may be difficult to keep people in the study due to the great mobility of the population. The greatest drawback, of course, is that it takes many years until all of the data can be collected and thus any conclusions reached. This type of study has yet to establish a link between toxicity and a particular toxicant.

Epidemiological studies, at best, allow the drawing of correlations between health effects and chemical exposures. Since they are uncontrolled, they cannot establish causation. Due to the drawbacks mentioned above, they rarely can even establish correlation. They are most useful in the rare situation where a unique toxic effect occurs in a population which has an unusual exposure. Two such cases are the occurrence of a rare liver cancer in vinyl chloride workers and the occurrence of a rare type of vaginal cancer in the daughters of women who took DES during pregnancy. Although epidemiological studies are limited in establishing toxicity, they may serve as checks on laboratory animal data. A particular effect established in exposed animals can be looked for in human populations similarly exposed. The presence or absence of such effects in people can provide additional evidence as to the toxicity of such chemicals in humans.

A summary of the characteristics of the four main types of toxicological assays is presented in Table 6.5. This clearly shows the differences in the types of testing done to assess different types of toxicity. It also suggests why it is so difficult to make comparisons among the different types of tests. This difficulty is reflected in problems of regulation and of communicating risk to the public.

Table 6.5   Characteristics of Some Assays Used in Human Risk Assessment

	Usual Subjects	Endpoint	Dose Criterion	Extrapolation Method Used
Acute Assay	Rodents	Death	50% lethality	Not generally done
Subacute Assay	Rodents	Multiple	Safe level	Safety factor
Chronic-Carcinogenesis Bioassay	Rodents	Cancer	Acceptable risk level	Mathematical modeling
Chronic-Epidemiology	Humans	Multiple	Variable	Not applicable

6.5   SHORT-TERM TESTS

In addition to the four main types of toxicity tests that have been mentioned—acute, subacute, and chronic studies on laboratory animals and epidemiological studies—one other type of test has been utilized extensively in recent years. This is the short-term in vitro, or test-tube, study. The most well known of these is the Ames assay, a test designed to check for mutagenicity, i.e., alterations in genetic material. This test is done with bacteria and can rapidly demonstrate whether or not a chemical can cause a certain type of genetic damage. However, the results in a simple test system are not easily transferable to multicelled organisms, much less humans. Thus it is not clear exactly how to apply these results, except in the qualitative sense that they indicate that a particular chemical has the potential to cause genetic damage. They are certainly not useful in establishing the dose at which such effects might occur in humans and cannot be applied to long-term exposures.

One way that these short-term tests have been used is as indicators or screening devices to determine which chemicals should be tested further. In particular, there seems to be some relationship between the ability of a chemical to cause genetic damage and its ability to cause cancer. Thus, chemicals which test positively in short-term assay are considered to be good candidates for long-term animal studies. However, there is dispute as to how firm this relationship between mutagenicity and carcinogenicity is, and the utility of short-term tests in this regard may undergo reappraisal as time goes by. However, there will be continuing work on developing such short-term studies because of the speed and relatively low cost of such tests. An additional impetus is the increasing influence of those who are against the use of animals in toxicology studies.

Overall, toxicity testing is a rather inexact science. It must be remembered that the aim is not to establish with certainty the doses at which effects will occur in humans, but rather to establish levels at which there is reasonable confidence that effects will not occur. Even this is a difficult goal, since it is not possible to experiment with the subjects of interest, humans, and scientists do not know enough to extrapolate with confidence from tests done on other organisms, mainly rodents. Furthermore, toxicity testing is done in the face of basic biological and chemical ignorance. If scientists understood exactly what happens to a foreign chemical once it enters the human body, then they could predict what effects it will have and thus obviate the need for any toxicity testing.

Because of the uncertainties involved in toxicity testing, there is a strong interaction between the tests themselves and the purposes for which the tests are performed. This interaction will be explored in the following chapter.

7

Risk Assessment and Risk Management

Chapter 6 provided an overview of toxicity assessment, also known sometimes as hazard assessment. This is the attempt to quantify the toxicity of a particular chemical. This type of assessment provides some measure of the inherent toxicity of the chemical to the exposed organism. To go one step further, i.e., to determine the risk to a particular human population from a toxic substance, requires knowledge of the exposure of that population. This determination is known as a risk assessment and is a combination of both toxicity assessment and exposure assessment.

7.1 EXPOSURE ASSESSMENT/DIRECT ENVIRONMENTAL MEASUREMENT

Measurement of exposure can be accomplished in a variety of ways. One is by direct measurement of the toxic substance in the medium to which the population is exposed. For example, analyses of concentrations in air or of levels in drinking water can be used to make estimates of the amounts of a chemical inhaled or ingested. However, since they generally reflect exposure at only one point in time, they are most useful in assessing risks of short-term, or acute, effects. In a few cases, mainly workplace exposures, analytical information over a period of years may be available and may be useful in assessment of chronic risks.

7.2   EXPOSURE ASSESSMENT/INDIRECT ENVIRONMENTAL MEASUREMENT

If direct measurements are not available, it is sometimes possible to make inferences from analyses which can be performed. For example, current workplace levels can be extrapolated backward in time if conditions are known to have been essentially identical for some period in the past. In this case, current levels are probably a good surrogate for previous levels. Of course, if industrial processes were changed, or new work rules went into effect, this type of inference would be much less valuable. In an environmental situation, it may be possible to trace the chemical of concern from its present location to a particular source and, from knowledge about this source and environmental movement of the chemical, again make some estimate of previous exposures.

For example, hydrogeological knowledge may be used to estimate the length of time it takes for a chemical to move from the surface down to the drinking water. Similarly, models of air transport or mixing in water may be utilized to provide some estimates of previous exposure levels. The further into the past the exposure of interest is from the time of the measurement, the greater the uncertainty that is introduced into the exposure assessment.

7.3   EXPOSURE ASSESSMENT/ANALYSIS OF EXPOSED POPULATION

A third way to assess exposure is to examine the individuals who are thought to have been exposed. For example, analyses of lead levels in human blood seem to give a good reflection of current exposure. Similarly, measurements of cholinesterase inhibition in blood provide good indicators of recent exposure to certain types of pesticides. These exposure indices are best suited to short-term risk assessments. When dealing with long-term toxicity, one fundamental problem is that many of the chemicals of interest are excreted by humans, and no trace remains soon after exposure has ceased. In such a case, even heavy exposure in the past may not be detected by current analyses. However, examination of current tissue or fluid levels may be of value when dealing with persistent chemicals, such as DDT or dioxin. In these situations, the chemicals are stored in the body for long periods and measurements of current body burdens provide some reflection of previous exposures. However, due to present uncertainties about half-lives and other characteristics, it is not possible to use these values to determine total previous exposure or length of exposure. Thus determinations of body burdens are useful mainly for measuring recent exposures and identifying exposed individuals rather than quantifying their exposure histories.

7.4 RISK ASSESSMENT

In addition to these fundamental problems in assessing long-term human exposure levels, there are difficulties in trying to apply toxicity assessments performed under specific laboratory conditions to human exposures that occur under real conditions. More specifically, carcinogenic risk assessments are generally done by exposing the laboratory animals to the same dose every day for a lifetime. This type of exposure never occurs in humans. Even in the workplace, exposure is interrupted during vacations, sick days, and when job changes occur. If someone is promoted from the line to a supervisory position, exposure may change radically. As uncertain as the situation is in the workplace, it is much easier to work with than environmental exposures. Even if levels of chemicals in drinking water are known, exposure is not at all constant, since water consumption changes seasonally, and people consume different percentages of liquid from their own supplies and from work or other environments at different times. In addition, it is rare for an individual to live in the same house, or even the same community for a whole lifetime.

Thus risk assessments are fraught with difficulties and it is not surprising that volumes are being written about the risk assessment process and ways that it might be improved. This is especially true with respect to carcinogenic risk assessment.

The uncertainties in this process lead to two different kinds of problems. One is the scientific question of how to arrive at the best value for risk, taking into account all of the uncertainty. The other is the societal one of how to best deal with risks when it is not possible to determine with great certainty the magnitude of the risk and how it might affect very specific populations.

7.5 RISK MANAGEMENT

This last task, dealing with known or perceived chemical risks, is known as risk management. Although it is often said that risk management starts with known risk assessments and then integrates social, economic, and political factors to arrive at a risk management strategy, this is overly simplistic. The history of risk management reveals that the risk management options that are chosen influence the risk assessments that are performed. They are not really independent. For example, the Delaney amendment to the Food, Drug and Cosmetic Act states that a substance can be used as a food additive only if it does not cause cancer in any species at any dose. This stipulation led to the use of tests that were designed to determine whether or not a chemical was a carcinogen rather than to quantitatively assess the chemical's carcinogenic potential. As time progressed, it was decided to manage chemicals other than food additives so that they would cause less than one cancer in 100,000 or 1,000,000 exposed individuals. Thus, an "acceptable risk" concept was introduced. This, in turn, led to modifications in the initial carcinogenesis assays to make them quantitative so they could be used to estimate the level at which each might cause this acceptable level of cancer.

Although risk management and risk assessment are really interactive, it is as profitable to discuss risk management separately as it was to deal with risk assessment separately. The fundamental aim of risk management is the reduction of exposure, since toxicity to humans is an inherent unchangeable property of each chemical. Exposure can be reduced in a variety of ways. In the workplace, this can be accomplished by modifications in the engineering or chemistry of the industrial process, or by utilization of personal protective equipment by the worker. An example of the latter is a respirator or a pair of impermeable gloves. In the general environment, exposure reduction may be accomplished by requiring treatment of effluents (gases or liquids) before they leave manufacturing facilities. Another method is to proscribe methods of waste disposal of toxic products. A rather drastic measure is to remove the population from the source of exposure. This approach has been rarely used, although it was adopted in the well-known cases of Love Canal in New York and Times Beach in Missouri.

In some cases, it has been decided that zero exposure is acceptable and that it is necessary to ban the production and/or use of the chemical. This has occurred most often with pesticides, such as DDT, dieldrin, and 2,4,5-T, but has also happened with some other substances such as the PCBs. Recently, the Environmental Protection Agency has proposed that the use of asbestos be phased out over the next decade. However, the number of banned chemicals is small and likely to remain so. Risk management of toxic chemicals is performed by the federal government under a large number of acts, with the major ones shown in Table 7.1.

Although the coverage and complexity of legislation have increased dramatically in the last two decades, there are relatively few substances for which quantitative, enforceable federal standards have been established. Appendix A provides a list of all of the chemicals for which maximum contaminant levels (MCLs) are now in effect nationally under the Safe Drinking Water Act. MCLs have recently been set for eight additional chemicals but these do not go into effect until January 9, 1989. Similarly, Appendix B provides an even shorter list of substances for which ambient air standards have been established under the Clean Air Act. Appendix C is a much longer list of allowable air levels of chemicals in the workplace. Even this long list represents only about 400 out of more than 50,000 chemicals in common use. It should also be noted that only a small number of chemicals have been added to this list since it was originally published in the early 1970s.

Table 7.1   Major Legislation Dealing with Regulation of Toxic Substances

Legislation	Types of Regulations
Clean Air Act	air pollution standards
Comprehensive Environmental Response, Compensation and Liability Act	remediation of hazardous waste sites
Dangerous Cargo Act	regulation of water shipment of hazardous materials
Federal Coal Mining Safety & Health Act	mine health and safety standards
Federal Food, Drug & Cosmetic Act	regulation of drugs, food additives, and cosmetics
Federal Hazardous Substances Act	regulation of hazardous household products
Federal Insecticide, Fungicide & Rodenticide Act	pesticide regulations
Federal Water Pollution Control Act	effluent and water quality standards
Hazardous Materials Transport Act	regulation of transport of hazardous materials
Lead-based Poisoning Prevention Act	regulations for lead control
Occupation Safety and Health Act	occupational health and safety standards
Resource Conservation & Recovery Act	hazardous waste management regulations
Safe Drinking Water Act	drinking water standards
Toxic Substance Control Act	hazardous substance regulations

There are other quantitative levels which have been set but they are much more limited in scope. An example is the tolerance level for a particular pesticide residue on a specific fruit or vegetable. On the other hand, many parts of the environment, such as indoor air, have yet to be regulated. An intense effort is underway to expand regulation, but, as should be evident from the first part of this book, it is impossible to generate appropriate data in a short time. The length of time and the amount of money needed for such studies, coupled with the deliberateness of the standard-setting process, excludes rapid response other than in emergencies.

7.5.1   Exposure Reduction

Since exposure reduction is the key to risk management, it is important to discuss how decisions are made regarding the extent of and methodology for exposure reduction that will be required. Unfortunately, there is no simple way to explain the way such decisions are made, because there are a variety of distinct agencies making these decisions and they must abide by a number of different legislative mandates. In the example cited previously, the agency in charge of enforcing the Delaney amendment, the Food and Drug Administration, has little leeway in deciding on risk management strategies in the case of carcinogens that are proposed for use as food additives. This same agency, however, has much more latitude when dealing with other toxic substances, especially drugs. In this case, the agency may balance the benefit of the chemical against its potential risk and may allow the production and marketing of drugs that are known to cause significant toxicity at levels near the effective dose. Some of the different standards that may be used by various agencies in regulating carcinogen follow:

FOOD AND DRUG ADMINISTRATION (USA)

[No] additive shall be deemed to be safe if it is found to induce cancer when ingested by man or animal, or if is it found, after tests which are appropriate for the evaluation of the safety of food additives, to induce cancer in man or animal. . . .

ENVIRONMENTAL PROTECTION AGENCY (USA)

The Administrator of EPA must weigh the risks to health from exposure to a carcinogen against the costs of controlling its use and the benefits of allowing continued use.*

*There are a few exceptions to this general rule.

OCCUPATIONAL SAFETY AND
HEALTH ADMINISTRATION (USA)

The Secretary . . . shall set the standard which most adequately assures, to the extent feasible, on the basis of the best available evidence, that no employee will suffer material impairment of health or functional capacity even if such employee has regular exposure to the hazard . . . for the period of his working life.

The Occupational Safety and Health Administration, which is responsible for setting acceptable standards for workplace exposure, operates under legislation that utilizes feasibility of exposure reduction as one of the criteria. In this situation, even a carcinogen can be allowed in the workplace as long as the levels are below a certain standard. Another agency that also utilizes feasibility in some of its risk management decisions is the Environmental Protection Agency. For example, requirements for controlling toxic chemicals in effluents are sometimes couched in terms of the best available control technology. In this case, the final level is not established, but rather the means that must be used to limit emissions to the environment.

7.5.2 Other Risk Management Strategies

There are many other possible ways to make risk management decisions, but there is not enough space in this brief volume to deal with all of them. Other examples might include the use of risk-risk decisionmaking where the risk of a chemical with known toxicity is compared to the risk of another substitute chemical about which less may be known. In addition, cost-benefit considerations may be taken into account as well as risk-benefit evaluations. That is, decisions might be made among alternatives on the basis of economic implications of various possible approaches, as well as the relative risk each poses.

7.6 RISK PERCEPTION

Both the regulations and the way that they are applied by the responsible agencies reflect public attitudes toward particular types of chemicals and specific kinds of risks. The way that people perceive risk is quite different from the way the scientist analyzes risk, and this dichotomy has led to conflicts in the public policy arena. It has been found that people tend to view catastrophic risks, e.g., airplane crashes, as greater than ordinary ones, e.g., automobile accidents. Similarly, man-made chemicals are viewed as riskier than natural ones; voluntary risks, e.g., smoking, as less significant than involuntary ones, e.g., air pollution; and those with immediate effects as less of a risk than those with delayed effects. Other factors have also been uncovered and together may contribute significantly to EPA priorities and actions with respect to toxic substances. A list of many of these factors is given in Table 7.2. As an example of the impact of risk perception, many scientists feel that the recent and continuing preoccupation with risk management of dioxins represents an exaggerated response to a risk which is of lesser importance than others which have received less attention.

Table 7.2   Public Perception of Risk

Criteria	Characteristics Perceived as Lower Risk	Characteristics Perceived as Higher Risk
origin	natural	man-made
volition	voluntary	involuntary
effect manifestation	immediate	delayed
severity (number of people affected per incident)	ordinary	catastrophic
controllability	controllable	uncontrollable
benefit	clear	unclear
familiarity	familiar	unfamiliar
exposure	continuous	occasional
necessity	necessary	luxury

7.7   ROLE OF CONGRESS

Before ending this section, it is important to note that the U.S. Congress has overruled the regulatory agencies in some cases and imposed its own risk management strategies. The most flagrant case is the nonregulation of the use of tobacco products, especially cigarettes. In this case, the Congress decided that the appropriate management strategy was to try to limit exposure by putting warning labels on the product. Another situation where an agency was overruled was the case of saccharin. As a food additive, it should have been banned under the Delaney amendment. However, the Congress has voted to exempt it from this provision and continues to renew this exemption. Again, in this extreme case, the Congress decided that labeling was the best risk management strategy.

Now that the basic principles of toxicology, toxicity assessment, risk assessment, and risk management have been presented, it is appropriate to examine the way in which a number of different toxic substances have been dealt with in the United States. These case histories will reveal how these principles are applied in practice and should give the reader a better appreciation of toxicology as a dynamic science which continually influences and is influenced by our society.

8

The Case of the Artificial Sweeteners

Many of the toxicological and risk assessment techniques that have been discussed have changed greatly during the past 20 years. The continuing saga of the development, marketing, and regulation of artificial sweeteners spans this period and provides a good introduction to the way in which the general principles have been applied in practice.

The ability of saccharin to impart a sweet taste to food and beverages was discovered nearly 100 years ago. Much more recently, in the late 1930s, it was found that cyclamates also shared this property. Both of these substances were proposed as food additives and went through the normal battery of testing that was commonplace before the 1960s. These were mainly acute and subacute tests on laboratory animals to determine if these chemicals caused toxic effects after single- or short-term exposures. They were not subjected to formalized testing protocols, although they were evaluated using generally accepted criteria for scientific validity that were followed by most scientists.

In 1958, the situation changed as the result of amendments to the Food, Drug and Cosmetic Act of 1938. The amendments changed the criteria for acceptibility of food additives, and specifically prohibited the use of any food additive that caused cancer at any level in any animal (the Delaney amendment). Since there were nearly a thousand food additives in use at that time, it was not thought feasible to perform a careful examination of each one. Instead, the list of additives was sent to scientists with known expertise in the toxicology of such substances and each was asked to comment on those that they thought to be unsafe or of questionable safety. This comment period resulted in the deletion of only six additives from the list, and so the vast majority of chemical additives were then approved without the necessity of further testing. These chemicals were considered "generally recognized as safe" (GRAS) and the list of these approved additives was known as the GRAS list.

Both saccharin and cyclamates were on this list. Neither was used very heavily at that time. The original use of these additives was in products aimed at diabetics who needed to control sugar intake. However, with the new social norms emphasizing a slim appearance, there was an increasing demand for low-calorie products, leading to expanded use of these noncaloric sweeteners, cyclamate and saccharin, that decreased the caloric content of beverages and food products. They were used together in sweeteners, since the combination had greater sweetening power and palatability than either alone.

8.1 CYCLAMATES

With the increasing demand for these artificial sweeteners, there was growing concern about possible adverse effects appearing in the general population. Some side effects might be considered acceptable for diabetics who should clearly benefit medically from use of these sweeteners, but toxic effects in the general populace would be less acceptable, since no commensurate medical benefit could be visualized. In view of this, the Food and Drug Administration (FDA), which has jurisdiction over food additives, asked the National Academy of Sciences to convene an expert panel to look into the safety of nonnutritive sweeteners.

The National Academy of Sciences (NAS) is chartered by the federal government, but it is an independent organization that elects its own members. It is made up of over a thousand prestigious scientists from a variety of scientific fields. The NAS performs most of its studies through its action arm, the National Research Council, and most of its funding derives from the organization that wants the evaluations performed. The actual studies are done under the auspices of the academy, but members of expert panels are made up of scientists with the special expertise required without regard to whether or not they are academy members.

The National Academy undertook reviews of saccharin and cyclamates in 1955, 1962, and 1968 and indicated a general satisfaction with the safety of these chemicals. It did indicate, however, that large intakes should be avoided, and also expressed concern about the increased usage of these materials.

In June 1969, not long after the last of these reviews, the producer of Sucaryl (10:1 cyclamate to saccharin), Abbott Laboratories, reported that bladder cancer had occurred in animals exposed for long periods of time to Sucaryl that was directly implanted into the bladder in a cholesterol pellet. This produced some concern, but it appeared that this route of exposure was so unlike normal human exposures by ingestion that the studies could not be applied to humans and the Delaney amendment could not properly be invoked.

Several months after this, in early October, Abbott reported that lifetime rat feeding studies of Sucaryl also revealed the presence of cancer in some of the exposed animals. At about the same time, studies by an FDA scientist, indicating that cyclamates produced birth defects when injected into chicken eggs, were reported on national television and provoked great public interest and anxiety. About a week after receiving the Abbott data, the FDA requested that the National Academy expert panel reconvene and examine the new findings, especially the carcinogenesis studies. One day after this meeting, the Secretary of the Department of Health, Education and Welfare (now the Department of Health and Human Services) announced that cyclamates were to be banned as food additives based on the Delaney amendment provisions.

Thus a clear risk management decision seemed to have been made. The material was banned. However, not all cyclamate-sweetened products were originally included, and even those which were banned were not considered dangerous enough to be recalled. This initial ban was followed by a period of about two years during which cyclamate-containing products were phased out under changing risk management decisions.

8.2 TOXICITY ASSESSMENTS AND RISK MANAGEMENT

Even more significant than this uncertainty with respect to the most appropriate risk management approach was the fundamental scientific question as to what the toxicity assessment showed. Remembering that the experimental animals who developed cancer were exposed to Sucaryl, a mixture of saccharin and cyclamates, how can it be concluded that only one component caused the cancer and that the study demonstrated which one this was? It could be argued that since Sucaryl is 10 parts cyclamates and only 1 part saccharin, the cyclamates are most likely the active agent. On the other hand, saccharin is a much more potent sweetener, so perhaps it is a much more potent carcinogen also. These, of course, are speculative statements and there is no scientific basis for banning one chemical, cyclamates, on the basis of a study of a mixture of two chemicals.

Thus the risk management decisions were not based entirely on the carcinogenicity assessment and the Delaney amendment. Other factors must also have been important. One possible contributor was the claim made that cyclamates are teratogens, i.e., they cause birth defects. However, even this claim was not well substantiated. Without inside knowledge, it is not possible to decide what weight this had and what role nonscientific factors played in the risk management decisions made at this time.

An interesting sidelight to this situation is that the data on Sucaryl and cancer were not made public at the time of the decision. They were published in February 1970 in *Science* magazine. Examination of the study protocol and results reveals that this research was of questionable validity in even establishing that the mixture was carcinogenic. There was no dose-response; cancer showed up only in the most highly dosed animals. In addition, the protocol was changed after

79 weeks. Thus the publication of the results raised questions as to whether or not the toxicity assessment was the crucial factor in the risk management decision that was ultimately made. A summary of the data published in this critical study is shown in Table 8.1.

Indeed, Abbott Laboratories has petitioned the Food and Drug Administration a number of times to allow cyclamates to be used in the United States again. The company contends that there is no good toxicological evidence that cyclamates do cause cancer. All of the petitions have been denied. It should be noted that cyclamates are used as food additives in other countries. Many countries followed the U.S. ban in 1969 with bans of their own based on information gained from the U.S. Food and Drug Administration. However, after the publication of the experimental evidence, several countries reevaluated the situation and allowed cyclamates back on the market. In some cases, this occurred in the early 1970s.

Table 8.1 Summary of Long-Term Feeding Study of Sucaryl

Group	Daily dose (mg/kg day) C/S[c]	CHA[d]	No. of animals alive at week[a] 0 M	F	104 M	F	No. tumors[b] M	F
A	0	0	35	45	13	26	0	0
B	500	25[e]	35	45	10	19	0	0
C	1120	56	35	45	8	23	0	0
D	2500	125	35	45	12	22	7	1

[a]Ten males and ten females died or were killed for interim study by the 56th week.
[b]Urinary bladder tumors as agreed upon by all of the pathologists on the basis of the slides available. Four to eight tumors were diagnosed as carcinomas by different pathologists.
[c]C/S = Sucaryl = 10:1 sodium cyclamate to sodium saccharin
[d]CHA = cyclohexylamine, a metabolite of cyclamate
[e]At week 79, groups B,C & D were divided into equal sized subgroups and one subgroup received additional CHA in the amount noted.

Source: Adapted from Price, J. M., Biava, C. G., Oser, B. L., Vogin, E. E., Steinfeld, J., and Ley, H. L., "Bladder Tumors in Rats Fed Cyclohexamine or High Doses of a Mixture of Cyclamate and Saccharin." *Science* 167, 1131–32 (1970).

8.3 SACCHARIN

Although the presence of saccharin in the mixture that was associated with cancer in rats was ignored in the risk management process, a number of scientists recognized that more research was needed to evaluate the carcinogenic potential of this chemical. A number of studies were undertaken utilizing a variety of test species. By 1972, several of these studies had been completed and the NAS was requested to review the situation and make recommendations.

This time, instead of a single meeting, there were a number of meetings and open publication of the findings of the academy committee in December 1974. The report revealed that saccharin showed very similar characteristics to Sucaryl when tested for carcinogenicity. Very high doses appeared to lead to bladder cancer, but cancer appeared only in the most highly exposed animals. Again questions were raised about the interpretation of these results, and the academy committee recommended that a number of additional steps be taken before a firm conclusion was drawn. These included investigations of possible infectious agents in the bladder, the possible formation of bladder stones at the very high doses administered, and a variety of other potentially complicating factors.

At about this same period, scientists in Canada were also looking into the carcinogenic potential of saccharin. They went one step further than the U.S. scientists did and looked at two generations. They fed saccharin to animals that were then mated, and they also fed saccharin to the offspring of these exposed animals. Although results were again equivocal in the first generation (the parents), there seemed to be a significant increase in cancer, linked to saccharin exposure, in the second generation.

On the basis of these studies, the FDA decided that the criteria mentioned in the Delaney amendment had been met, and it took steps to ban saccharin as a food additive. That is, the FDA took the risk management approach to saccharin which it had previously taken for cyclamates. The ban was proposed in 1977, eight years after the cyclamate ban.

8.4 CONGRESSIONAL INTERVENTION

At this point, however, the situation changed from what had occurred eight years before. The Congress of the United States decided that this ban was not warranted and passed legislation temporarily extending the period during which saccharin could still be used. This was originally set at three years, but similar legislation has been passed twice since then preventing a ban on saccharin from taking effect. It remains on the market today.

Thus, of two artificial sweeteners that were in use 20 years ago, one has been banned and the other is still on the market. Interestingly, the one for which there is the strongest evidence of carcinogenic potential is the one that is still on the market. Indeed, a report published by the NAS in 1985 concludes that cyclamate by itself is not carcinogenic. This provides a clear illustration of the importance of nonscientific factors in the risk management process. A chronology of cyclamate and saccharin regulation in the United States is provided in Table 8.2.

Another aspect of this situation bears mentioning. During the 1970s, there was a strong push toward determining the levels at which chemicals can cause cancer. Thus the studies that were originally designed to provide yes-no answers in response to the Delaney amendment were modified to provide quantitative measures of carcinogenic potential in response to other legislation governing toxic substances. When this was done, it was recognized that saccharin is a very weak carcinogen in laboratory animals and, if these animal studies are accepted as applicable to humans, a very weak carcinogen in humans.

This conclusion has been supported by a number of epidemiological studies which have been performed. In none of these has it been possible to conclusively link increased bladder

Table 8.2 Chronology of Cyclamate and Saccharin Regulation in United States

	Cyclamate	Saccharin
1897		First made
1912		First used in U.S.
1937	First discovered	
1950	Accepted as drug intended for use in food and beverages by diabetics and others who must restrict their intake of sugar	
1959	Reclassified as food additive on GRAS* list	Accepted as GRAS food additive
1969	Removed from GRAS list	
1970	All foods and drugs containing cyclamates off the market	
1972		Saccharin restricted to current uses and removed from GRAS list
1977		Ban on saccharin proposed by FDA. Congress voted 18-month moratorium pending further study. Moratorium renewed on bi-yearly basis.
1987		End of current moratorium period

*GRAS = Generally Recognized As Safe.

cancer with consumption of cyclamates or saccharin. Of course, one population that may consume the greatest amounts, diabetics, shows a number of toxic effects from the disease, and this complicates the study of possible effects due to saccharin. However, it does appear clear that the cyclamates or saccharin are, at worst, very weak human carcinogens.

8.5 ASPARTAME/ANOTHER ALTERNATIVE

With the banning of cyclamates and the possibility of a ban of saccharin, an intensive effort to find alternative sweeteners was undertaken in the 1970s. The sweetener which was the most successful result of this work is aspartame. This was discovered in 1965, proposed to the FDA as a sweetener in 1973, and approved for use in 1981 for dry foods, in 1983 for carbonated beverages, and in 1986 for still other uses. Laboratory investigations of aspartame have not revealed any significant toxic effects at levels corresponding to those likely to be ingested by humans. A carcinogenesis bioassay in rodents was performed and has been interpreted by some to suggest that aspartame causes brain tumors. However, the consensus of scientific opinion is that the data do not support such a conclusion.

There have been a number of claims that aspartame has caused adverse effects in humans. It has been suggested that it induces seizures in susceptible individuals, causes headaches and allergic reactions, and can alter the functioning of the nervous system. However, these are based on individual reports and there has been no clear pattern of complaints. No epidemiological studies have been performed. Thus there is no valid scientific evidence to support a connection between aspartame and any of the reported effects at this time. It is possible, however, that significant side effects will be found after aspartame has been used by more people and for a longer period of time.

The approval of aspartame is a good example of one trade-off that has sometimes been made in risk management: acceptance of a compound with no use experience, but little animal evidence of toxicity, in preference to one with long use experience, but some evidence of toxicity on the basis of animal studies. In the present example, cyclamates remained banned despite long-term human evidence of safe use, while aspartame was approved based on animal toxicity data.

8.6   A NEW RISK MANAGEMENT APPROACH

In the past year, the Food and Drug Administration has proposed that it take.into account differences in carcinogenic potency in food additives rather than just determining whether or not an additive is a carcinogen. This approach is based on the *de minimis* principle, or the principle of negligible harm. The FDA is arguing that the carcinogenic potency of some chemicals is so low that it has no real significance in the human population. While theoretically an individual might be affected, the probability of this happening is small and not worthy of further action. This principle was spelled out by the FDA in the case of methylene chloride as follows:

> "Because methylene chloride has been shown at a statistically significant level to be a carcinogen by inhalation in the NTP mouse bioassay, if the Delaney anticancer clause (21 U.S.C. 348(c)(3)(A)) is to be interpreted as applying even if a *de minimis* risk is involved, FDA could not find that use of methylene chloride for decaffeinating coffee is safe. Yet, if the associated risk is essentially negligible, there is no gain to the public, and the statutory purpose is not implemented, if the words of the statute are interpreted not to leave the agency any discretion to apply it reasonably. The calculated risk for this use of methylene chloride is extremely low. The risk (no greater than 1 in 1 million and probably closer to 1 in 100 million) is so low as to be essentially nonexistent. Given such a low level of risk, FDA has concluded that there would be no safety gain to the public if it interpreted the Delaney Clause to require a ban on this use of methylene chloride. Therefore, FDA, exercising its inherent authority under the *de minimis* doctrine, concludes that the Delaney Clause does not require a ban in this situation. Because there are no other known safety problems with this use of methylene chloride, FDA finds that the use of methylene chloride to decaffeinate coffee is safe." [*Federal Register* 50, 51555 (December 18, 1985).]

If this principle is adopted, it can be seen that the latitude for making risk management decisions with respect to food additives will be broadened. However, as the case of artificial sweeteners indicates, significant latitude already exists. In summary, examination of the way the most stringent federal risk management regulation, the Delaney amendment, has been applied in practice reveals the complex interaction of toxicity assessment, risk assessment, and risk management in the United States.

9

The Case of Asbestos

In the previous chapter, the chemicals of concern were food additives, and the potentially exposed group represented a large part of the total population of the United States. In addition, the chemicals were evaluated mainly on the basis of animal experimentation—in particular, the rodent bioassay. In this chapter, another situation will be explored, one that represents quite different risk assessment studies and risk management decisions.

9.1 ASBESTOS AND ASBESTOSIS

Asbestos has been in common use for about 100 years. It was first developed commercially in England, and the mining and fabrication of this natural material grew steadily at the beginning of this century. It was not long, however, until the suspicion arose that asbestos was causing adverse health effects in workers. Since the effects were on the respiratory system, they were not easily identified against a high background incidence of other respiratory problems, especially tuberculosis and silicosis, at that time. By the early 1930s, however, it was clear that asbestos was causing significant losses in lung function among workers and was contributing to early death in some of the heavily exposed individuals. This lung ailment was called asbestosis.

Since a number of years of exposure were required before asbestosis became evident, it was decided that lowering the dose would decrease the incidence of the disease. This decrease could result either from lowering the dose below the toxic level or extending the number of years of exposure needed for the effect to appear. If this time could be increased past a normal life span, the effects would essentially be eliminated. Thus, the risk management approach that was taken was to decrease worker exposure. This was accomplished by a combination of better processing methods and greater use of personal protective devices, such as respirators.

Since asbestos seems to exert its toxic effects when inhaled, the usual experimental exposures of laboratory animals through the food are not appropriate. Studies which involve air exposure are difficult to design, and inhalation facilities are difficult and expensive to construct. Thus, inhalation toxicity studies were not done on a regular basis until fairly recently. In addition to the technical problems, rats are obligatory nose breathers and so might be expected to react differently than humans to inhaled substances. In fact, when studies were performed recently to test this idea, it was found that the structural changes in the rat lung after asbestos exposure were quite different from those in humans who had been similarly exposed. In summary, animal models have not been and still are not readily available for assessing human toxicity of asbestos.

In the absence of such controlled animal experimentation, it was difficult to decide what an acceptable level of exposure might be. The basic risk management approach was to reduce exposure as much as seemed reasonable on the basis of previous experience and feasible in terms of costs and then see what happened. This was the situation until 1946, when a more detailed examination of the human data was available. On the basis of this, a nongovernment group called the American Conference of Governmental Industrial Hygienists (AC-GIH) recommended a limit of 5 million asbestos particles per cubic foot of air averaged over an 8-hour work day.

9.2   ASBESTOS AND LUNG CANCER

One of the precipitating factors leading to the recommendation for such a level was the suggestion in 1941 that asbestos caused lung cancer as well as asbestosis. At this time it was discovered that lung cancer occurred in 10% to 20% of autopsied asbestos workers. During the next 10 years, confirmation of this initial suggestion came from further epidemiological investigations in England, the Soviet Union, and Japan, as well as the United States. As with asbestosis, there was a long latent period between the beginning of exposure and the appearance of the cancer, in this case about 25 to 30 years.

It thus appeared that the initial risk management decision to lower the exposure had, indeed, reduced the incidence of asbestosis but had not been stringent enough to prevent another previously unknown adverse health effect, lung cancer. Here again, the absence of good animal experimentation was crucial in the risk management decisions which were made. Epidemiological studies were not a good substitute, since they lacked sensitivity and were confounded by other toxic effects that affect the same organ.

9.3   ASBESTOS AND MESOTHELIOMA

Since the limit recommended in 1946 was based on the suspicion that asbestos caused lung cancer, no further decreases were recommended when this suspicion was seemingly confirmed. However, in the 1960s asbestos was gradually linked to a third type of life-threatening toxic effect, mesothelioma (cancer of the lining of the lung). This was very difficult to detect in exposed individuals, because it occurred at such low incidence and because it took so long to appear. The incidence, based on autopsies, is less than 0.1% of the exposed population, and the time between exposure and disease onset is 35 to 40 years. It is also difficult to show any relationship between intensity or duration of exposure and incidence of the disease. Last, routine animal experiments are not appropriate due to the difficulties mentioned earlier. It has, however, been possible to produce mesothelioma in rodents when the asbestos has been injected into the area where the cancer develops.

9.4   RISK MANAGEMENT OF ASBESTOS

With the growing evidence of multiple effects from asbestos and the impossibility of confidently establishing a no-effect level, there has been a gradual move toward establishing lower and lower workplace limits. A history of occupational standards in the United States is given in Table 9.1. In addition, the recognition that even low levels of exposure can lead to adverse effects has led to concern about the general population. In response to this, the Environmental Protection Agency acted against industries which were emitting asbestos into the atmosphere. Emission limits were promulgated in 1973. In addition, surfacing of highways with asbestos tailings was prohibited. Another federal agency, the Consumer Product Safety Commission (CPSC), was concerned about the use of asbestos in consumer products. It acted in 1978 to ban the use of asbestos-containing patching compounds and artificial em-

Table 9.1   History of Development of Standards for Occupational Exposure to Asbestos in United States

Year	Standard
1946	ACGIH—5 mppcf
1968	ACGIH—2 mppcf—12 f/ml
1970	ACGIH standard of 12 f/ml adopted by U.S. Secretary of Labor as interim standard
1971	OSHA issues temporary emergency standard—5 f/ml
1972	OSHA makes 5 f/ml permanent standard
1975	OSHA recommends standard of 0.5 f/ml
1976	OSHA adopts 2 f/ml standard
1979	NIOSH/OSHA committee recommends standard of 0.1 f/ml
1980	ACGIH—2 f/ml—chrysotile; 0.5 f/ml—amosite; 0.2 f/ml—crocidolite
1986	OSHA adopts 0.2 f/ml standard with a 0.1 f/ml action level

Note: For regulatory purposes, fibers have been defined as particles having an aspect ratio of 3:1 and length > 5 μm.

mppcf = million particles per cubic foot
f/ml = fibers per milliter
ACGIH = American Council of Government Industrial Hygienists
OSHA = Occupational Safety and Health Administration
NIOSH = National Institute of Occupational Safety and Health

berizing materials. In addition, in 1979, the CPSC negotiated a voluntary agreement with hair dryer manufacturers for the removal of asbestos from this product. Thus risk management of asbestos has involved many possible sources of exposure and a variety of management approaches.

More recently, the EPA took action to encourage management of possible asbestos exposure in schools. It required schools to inspect their buildings for asbestos, especially friable (crumbly) asbestos, and to report their findings to the staff and the parents of the children. While this has encouraged actions designed to deal with any friable asbestos that is found, the steps that have been taken have varied widely from school district to school district. This illustrates a rather rare occurrence: risk management decisionmaking on a local basis for a chemical of great national concern.

On a global scale there are also differences in risk management of occupational exposure to asbestos. A summary of the limits in other countries (Table 9.2) illustrates this diversity.

This unusual situation may be short-lived. The difficulty of ascertaining any no-effect level and the ubiquitous human exposure has led in recent months to the proposal that the use of asbestos be phased out over the next decade. Thus the ultimate risk management step, a ban, may yet be taken. It should be emphasized that such a step will not eliminate all subsequent exposure. Asbestos is present in a large percentage of all structures in the United States and until all of these have been destroyed, it is likely to remain. Even after that, asbestos will remain scattered around the earth and continue to be moved by wind or water. Since it is a mineral, it will not degrade easily, and the main hope is that it will eventually be buried and, in essence, go back to where it originated.

9.5   ASBESTOS AND SMOKING

There is a sidelight to this story that brings up an issue which was not dealt with in the brief review of toxicity and toxicity assessment. How do toxicologists deal with the question of interactions among chemicals? Theoretically, chemicals' effects

Table 9.2   Occupational Exposure Limits for Asbestos

Country	Limit (f/ml)
Australia	Chrysotile, amosite—1; crocidolite—0.1 (4 hours)
Austria	1
Belgium	Crocidolite—0.2; other—2 (4 hours)
Canada	2 (8 hours)
Denmark	Crocidolite—0.1; other—1
Finland	2
France	2—breathing zone; 1—workroom air (working day)
Germany (D.R.G.)	2
Germany (F.R.G.)	1
Ireland (Republic of)	Crocidolite—0.2; other—2
Israel	1
Italy	2
Japan	Crocidolite—0.2; other—2
Netherlands	Crocidolite—0.2; other—2 (4 hours)
New Zealand	Crocidolite—0.2; other—1
Norway	2 (15 minutes)
Singapore	2
South Africa	5 (mines); 2 manufacturing
Sweden	0.5
Switzerland	2 (8 hours)
U.K.	Crocidolite—0.2; chrysotile—1; amosite—0.5 (4 hours)
U.S.	2 (8 hours); 10 (15 minutes)
U.S.S.R.	Asbestos-containing dust—8 mg/m^3; Amphiboles and A/C dust—5–6 mg/m^3; Fine dust—mg/m^3

Note: Crocidolite, amosite and chrysotile are different forms of asbestos.

Source: Adapted from Meek, M.E., H.S. Shannon and P. Toft, "Case Study—Asbestos" in *Toxicological Risk Assessment* (Vol. II), D.B. Clayson, D. Krewski, and I. Munro, eds., CRC Press, Boca Raton, Florida, 1985.

could be simply additive, could be less than additive, or more than additive. There appear to be examples that illustrate each of these possibilities. Asbestos provides an illustration of the greater-than-additive possibility when asbestos exposure is coupled with exposure to tobacco smoke. It has been found that exposure to asbestos increases the probability of developing lung cancer by 5 times over background levels, while smoking increases the risk by about 12 times. However, if an individual who smokes is exposed to asbestos, the total risk appears to be multiplicative. This individual is 60 times more likely to get lung cancer than the nonsmoking, non-asbestos-exposed individual. This type of interaction is found in several other workplace exposures, where smoking also appears to magnify the toxicity of another chemical in greater than an additive manner.

This brings up another question. If lung cancer incidence in asbestos workers can be reduced by limiting worker exposure to tobacco smoke, can this be considered a risk management tool? If the worker is aware of this, cessation of smoking certainly could be used by the individual as a voluntary risk management technique. It is not clear, however, whether or not this could be required of workers. Thus the use of smoking cessation to reduce lung cancer of asbestos workers may be a limited approach.

Another issue that is raised by the interaction is how to divide responsibility for the causation of lung cancer. This is not just an academic question, but is of importance in litigation where an injured party, the individual with the cancer, brings suit to collect damages from the responsible party or parties. If smoking is a contributory factor, does the individual or tobacco company bear some responsibility for the lung cancer? If so, how can the responsibility be apportioned? What percentage of the liability is the employer responsible for? These are examples of a number of similar questions which are currently being decided in the courts. The results of these court cases may well have an impact on future risk management approaches.

Thus the case of asbestos illustrates a situation where animal experimentation is not an effective tool and where human epidemiology is the technique by which the toxicity of asbestos was revealed. The weakness of this technique is evident in the long time it took to discover that this substance causes asbestosis, then that it causes lung cancer, and last that it appears to cause mesothelioma. In addition, the lack of data from controlled animal experimentation has made it impossible to determine a reasonable no-effect level. This in turn has led to problems in establishing consensus risk management approaches and, finally, to the proposed banning of the material.

This case also illustrates the differences in approach that can occur once a toxicant is suspected of being a general population hazard as well as an occupational threat. Based on the cyclamate case and other historical examples, it appears to be much more difficult to ban a substance which is only a workplace hazard as opposed to one which can affect everyone. This seems to hold for carcinogens as well as acute toxicants.

10

The Case of Formaldehyde

In the previous two chapters, the history of efforts to manage two different types of chemicals was discussed. One type was represented by synthetic chemicals used as food additives and the other by a naturally occurring material used in a variety of products, especially building materials. In this chapter, a slightly different type of compound will be discussed: formaldehyde.

Formaldehyde differs from the previous examples in at least one essential: it not only occurs in the natural environment but also is found naturally in the human body. Therefore, it is a substance for which it might be assumed there would be little toxicological effect. Unfortunately, this assumption has turned out to be incorrect.

Formaldehyde has been used in commercial products since the early part of the twentieth century. Its use has grown and it is now widely utilized for the production of plastics and plywood as well as a host of other products. It is well known as a preservative and was utilized quite extensively as part of a building insulation material until recently. It is produced from automobile exhaust and is also a component of cigarette smoke. Since it is so widely used and produced, it can be found in both the general environment and the workplace.

10.1 ACUTE EFFECTS OF FORMALDEHYDE

As with other materials used in manufacturing, workplaces generally represent the highest exposure situations, and it was in industry where the risk of formaldehyde inhalation was first discovered and managed. It was found that people who are exposed to formaldehyde at high enough levels show eye, respiratory, and skin irritation and, as levels increase, exhibit more severe effects such as nausea, dizziness, and vomiting. These are acute effects and so were easy to link to the responsible agent, formaldehyde.

As a result of this linkage, the ACGIH recommended a 10-ppm threshold limit value for workplace air in 1946. This represented the maximum allowable time-weighted average concentration over an 8-hour day. This was lowered two years later to 5 ppm and, in 1973, to 2 ppm. The continued lowering of the recommended maximum exposure levels was due to persistence of the irritant effects in workers despite the decreasing limits. In 1971, the Occupational Safety and Health Administration (OSHA) set a threshold limit value of 3 ppm based on recommendations made by another nongovernmental group, the American National Standards Institute. This 3-ppm standard is still in effect. However, it should be noted that in 1976 the National Institute of Occupational Safety and Health, the scientific arm of OSHA, recommended that the permissible exposure limit be reduced to 1 ppm. This recommendation was not accepted by OSHA.

10.2   FORMALDEHYDE AND CANCER

The risk management climate changed dramatically in 1979 as a result of reports that formaldehyde caused nasal cancer when inhaled by rats at concentrations of 14.3 ppm and 5.6 ppm in air. Subsequent animal studies seemed consistent with this one, and it was soon concluded that formaldehyde should be considered a carcinogen in rodents. However, epidemiological studies on workers in a variety of industries failed to show comparable effects. There was no convincing evidence that formaldehyde caused cancer in humans. However, it must be remembered that the general regulatory posture in the United States has been that carcinogenicity in animals is presumptive evidence of a similar effect in humans. This was the basis for the ban on cyclamates even in the absence of any confirmatory evidence in humans.

10.3   UREA-FORMALDEHYDE FOAM INSULATION (UFFI)

At about this same time, it was recognized that a variety of building products emitted formaldehyde and that the general population was continually being exposed to this material. This situation was exacerbated by the energy crisis that led people to reduce ventilation in their homes to reduce fuel costs. Tighter homes led to increased levels of indoor air contaminants. Of particular concern was the large number of people who appeared to show irritant effects in homes that were insulated with urea-formaldehyde foam insulation (UFFI).

As a result of this concern, UFFI was banned in a number of places in 1980. The Canadian government enacted a ban and one state, Massachusetts, also prohibited its use. This is a good illustration of different government units utilizing the same scientific data but making different risk management decisions. The Massachusetts decision reflects a fairly common occurrence in the United States: differing regulation in different states. Incidentally, this ban was overturned in court in 1982.

The overturning of this state prohibition occurred just before action at the federal level was taken; the Consumer Product Safety Commission (CPSC) voted to ban the use of UFFI in homes in 1982. This ban was based not only on the irritant effects of formaldehyde but also on the possibility of cancer developing in exposed individuals. The CPSC calculated that the risk could be as high as 50 cancers per million people exposed continually to UFFI. This is clearly greater than the "acceptable level" of one per million.

The ban was immediately challenged by industry and, after a lengthy court battle, was overturned. The court made this decision based on its estimation of the validity of the exposure information used in the risk assessment and also the validity of the studies done on rats that indicated formaldehyde is a carcinogen. Although the ban was overturned, the wide publicity which the original ban was given led to significant decreases in the use of UFFI. Indeed, a small industry sprung up consisting of people who claimed to be experts in removing UFFI from homes in which it had been installed.

10.4   RISK MANAGEMENT OF OCCUPATIONAL EXPOSURE TO FORMALDEHYDE

In the meantime, in 1983, the ACGIH added formaldehyde to its list of substances suspected as potential human carcinogens and lowered the recommended threshold limit value to 1 ppm. It also set a recommended ceiling limit of 2 ppm. This is the maximum allowable over any 15-minute period. Two years later, the Department of Housing and Urban Development promulgated regulations requiring that plywood and particleboard emit no more than 0.2 ppm and 0.3 ppm of formaldehyde, respectively.

All during this process of changing recommendations and regulations, there were meetings of numerous scientific panels to evaluate the research which was emerging. The consensus was that the evidence did confirm that formaldehyde is an animal carcinogen, but that clear evidence of human carcinogenicity was not available. In view of the uncertainties of extrapolation from the rodent data, it was not possible to reach consensus on the quantitative cancer risk that might exist in human populations.

Also during this time, there was pressure on the Environmental Protection Agency and the Occupational Safety and Health Administration to act. In December of 1985, OSHA finally proposed new standards for formaldehyde. This was done to meet the deadline set by the U.S. Court of Appeals in response to petitions from labor. Instead of a single standard, OSHA proposed two possibilities. One is to lower the permissible exposure limit to 1 ppm with an action level at 0.5 ppm. This is based on formaldehyde's carcinogenicity. The alternative is to lower the limit to 1.5 ppm with an action level of 0.75 ppm. This action level provides a value below which employers do not have to meet the standard's requirements for engineering control, protective clothing, monitoring, or medical surveillance. Irritant properties are the basis for this higher standard. It should be noted that in neither case is a short-term ceiling level proposed.

At present, hearings are underway on this proposal and it is not clear what the final standard will be. It does appear, however, that the standard will be lowered. Once OSHA took action, EPA was no longer responsible for addressing potential risks from occupational exposure. It thus terminated its investigation of this aspect of formaldehyde's adverse effects. The EPA will continue investigations of possible regulation needed to control general exposure resulting from the use of pressed wood products.

Although the end is not in sight regarding the ultimate risk management steps that will be taken with respect to formaldehyde, it is still possible to make comparisons with the previous examples. As with asbestos, the substance is one that results not only in workplace exposure but also general population exposure. It is thus potentially subject to management in a number of different arenas. As with asbestos, occupational exposure has been addressed by a continually shifting set of regulatory requirements, each of which involved lowering of acceptable limits. In contrast to asbestos, these changes have resulted in the main not from the discovery of new toxic effects, but from the findings that the established toxic effects could occur at lower and lower doses. A chronology of formaldehyde regulation in the United States is provided in Table 10.1. Table 10.2 again illustrates how such regulations can vary from country to country.

Table 10.1   Chronology of Formaldehyde Limits in United States

1946	ACGIH	10 ppm (TLV)
1948	ACGIH	5 ppm (TLV)
1963	ACGIH	5 ppm (Ceiling)
1971	OSHA	3 ppm (TLV), 5 ppm (PEL)
1973	ACGIH	2 ppm (Ceiling)
1976	NIOSH	Proposed 1 ppm (PEL)
1982	CPSC	Ban on UFFI
1983		Ban overturned in Court of Appeals
1983	ACGIH	1 ppm (TLV), 2 ppm (STEL)
1985	HUD	Plywood and particleboard cannot emit greater than 0.2 ppm and 0.3 ppm, respectively
1985	OSHA	Proposed 1 or 1.5 ppm (PEL); 0.5 or 0.75 ppm (action level)

TLV = Threshold Limit Value; PEL = Permissible Exposure Limit; STEL = Short Term Exposure Limit; CPSC = Consumer Product Safety Commission; HUD = Department of Housing and Urban Development.

Table 10.2   1976 Occupational Standards for Formaldehyde

Country	Standard ppm	Type
United States:		
Federal Standard	3	TWA (time weighted average)
	5	Ceiling
	10	30-min ceiling
ACGIH TLV	2	Ceiling
Florida	5	Ceiling
Hawaii	10	Ceiling
Massachusetts	3	Ceiling
Mississippi	5	Ceiling
Pennsylvania	5	TWA
	5	5-min ceiling
South Carolina	5	Ceiling
Bulgaria	4	Ceiling
Czechoslovakia	4	Ceiling (10 min)
Federal Republic of Germany	5	Ceiling
Finland	5	Ceiling
German Democratic Republic	4	Ceiling
Great Britain	10	Ceiling
Hungary	1	Ceiling
Italy	4	Ceiling
Japan	5	Ceiling
Poland	1.5	Ceiling
Rumania	2.5	Ceiling
UAR	20	Ceiling
USSR	0.4	Ceiling
Yugoslavia	5	Ceiling

Source: Adapted from NIOSH Criteria for a Recommended Standard. Occupational Exposure to Formaldehyde, DHEW Publication No. 77-126, U.S. Government Printing Office, 1976.

The conclusion that formaldehyde is a carcinogen has led to great changes in risk management procedures. This is not too surprising in light of the examples discussed in the previous two chapters. However, this conclusion has not led to a ban or a proposed ban of this chemical, as was the case for asbestos and cyclamates. One distinction that might contribute to the differences in regulation between formaldehyde and asbestos is that the latter has clearly been shown to be a human carcinogen while the former has not. Of course, cyclamates were not shown to be human carcinogens either, but regulation of these chemicals occurred at an earlier phase in the continually developing risk management strategy in this country.

11

The Case of Benzene

In the previous three chapters, chemicals which showed different toxicological properties and were utilized in quite different ways were managed in significantly different fashions. Interestingly, one which is accepted to be a human carcinogen has continued in widespread use while another substance, about which there is doubt concerning its carcinogenicity in any species, has been unavailable due to a ban on its use. Before trying to draw any conclusions about this disparity, it would be profitable to examine the risk assessment and risk management of a fourth compound: benzene.

Benzene is a component of a large number of products. These include naturally occurring substances, such as natural gas and crude oil, as well as a number of products of the petrochemical and petroleum industry. In addition, benzene is produced for use in a variety of industries such as printing, rubber fabricating, varnish, and adhesives. Benzene is also widely used as a solvent in chemical laboratories. Gasoline is probably the most familiar, widely used product containing benzene.

11.1 BENZENE AND ANEMIA

This chemical has been in commercial use for about 100 years, and its ability to cause toxic effects was suspected at the beginning of this century. It was noted that worker exposure to benzene was related to a variety of blood-related disorders, such as anemia. This led to the suggestion that benzene's toxicity is due to effects on the blood-forming cells in the body. In response to the growing evidence of its effects, an exposure limit of 100 ppm was proposed in 1927. When the ACGIH published its initial set of recommendations in 1946, this 100-ppm standard was adopted. However, this recommended limit rapidly decreased in the following two years, to 50 ppm in 1947 and to 35 ppm in 1948. This lowest concentration corresponded to that set by the state of Massachusetts on the basis of studies on toxic effects of benzene on workers in that state. In 1963, as more evidence accumulated about blood effects and exposure levels, the ACGIH proposed an even lower limit of 25 ppm. In the early 1970s, this recommended value was reduced to 10 ppm by the ACGIH as well as the American National Standards Institute. OSHA adopted this standard along with others during the early 1970s.

11.2   BENZENE AND CANCER

At the time the first health effects were noted in the United States, around 1900, there was some suggestion that benzene was also responsible for leukemia, a cancer of the blood. No formal studies were undertaken to investigate this possibility until the early to mid-1970s when a number of investigators examined information about workers in a number of industries which utilized benzene. From the patterns of exposure and toxicity in this country and abroad, it was concluded that there was link between benzene and leukemia. This, then, represents another of the limited number of situations where epidemiology has provided strong evidence of a link between a chemical and a specific chronic toxic effect.

As a result of the accumulating evidence implicating benzene as a human carcinogen, OSHA issued an Emergency Temporary Standard of 1 ppm in 1977. However, this standard never took effect because of stays that were granted by the court. OSHA promulgated a permanent standard a year later. The rationale behind the 1-ppm level proposed at that time was that it was the lowest feasible level and would therefore reduce the incidence of cancer to the greatest extent possible. This standard was also challenged in the courts.

After a series of lower court decisions, the standard came before the U.S. Supreme Court, which upheld the lower court decision to vacate the standard. The basic argument of the court, at least the one-vote majority of the court, was that OSHA had determined neither that a significant risk existed at the current standard nor that the new standard would significantly reduce this risk. In other words, the epidemiological data was not good enough to determine the exact exposure levels of the affected individuals and thus could not be used to determine the risk at a given level of exposure. In the opinion of the majority, it was not enough just to show that benzene caused cancer; it was necessary to show that it caused an unacceptable incidence of cancer above 1 ppm. This decision was handed down in 1980, some three years after OSHA had originally proposed the lower standard.

During the period of the court battles, carcinogenicity studies were performed in a few different laboratories on both rats and mice. These studies demonstrated that benzene caused a variety of different kinds of cancer in these laboratory animals, including leukemias. Feeding and inhalation studies were performed on benzene, and the results were similar with each route of exposure. In addition, other studies on experimental animals reproduced other effects on blood found in humans and also indicated that benzene could cause damage to genetic material. The carcinogenesis assays were of greatest importance to OSHA since they provided a quantitative set of data which could be used to calculate the cancer risk at specified levels of exposure and also provided convincing evidence that benzene was, indeed, the causative agent of leukemia in the workplace.

11.3   BENZENE REGULATIONS IN THE WORKPLACE

In 1983, on the basis of the accumulating evidence, a petition was sent to OSHA requesting an Emergency Temporary Standard. OSHA denied this petition. At the end of 1984, after some attempts at a negotiated settlement among industry, trade unions, and other interested parties were unsuccessful, a petition was filed with the Court to direct OSHA to proceed with benzene rulemaking on an expedited basis. In December 1985, OSHA proposed a standard for benzene that was essentially the same as that proposed in 1977. The exposure limit is proposed to be 1 ppm with an action level of 0.5 ppm.

In the proposal for the standard, OSHA provided quantitative risk assessments based on the epidemiological data that had been accumulated as well as from the controlled laboratory experiments on rodents. It thus appears that the rationale for the proposal meets previous court objections for specificity. However, there is a comment period which is required by law, and the possibility exists that OSHA will make some modifications in the standard as a result of the input received during this time. There is also the chance that legal action will result in further delays. In any event, there is no new standard at present, nearly 10 years after an initial proposal for such a standard was made. The chronology of regulatory effects to control occupational exposure to benzene is summarized in Table 11.1.

Table 11.1 Chronology of Benzene Limits in United States

1927		100 ppm limit proposed
1940s	State of Massachusetts	35 ppm
1946	ACGIH	100 ppm (TLV)
1947	ACGIH	50 ppm (TLV)
1948	ACGIH	35 ppm (TLV)
1963	ACGIH	25 ppm (TLV)
1971	OSHA	10 ppm (PEL)
1974	ACGIH	10 ppm
1978	OSHA	Proposed 1 ppm (PEL)
1978		Standard vacated by court
1985	OSHA	Proposed 1 ppm (PEL); 0.5 ppm (action level)

11.4   BENZENE AND AIR POLLUTION

At the same time that action on occupational exposure to benzene was being developed, other potential sources of exposure were also being addressed. Benzene was designated a hazardous air pollutant by the Environmental Protection Agency, and, in June 1984, emission standards were promulgated for a number of industries using or producing benzene. These are not quantitative standards which directly regulate the levels of benzene in outdoor air, similar to occupational levels for workplace air, but regulations to limit the amount of benzene emitted by particular facilities, especially petroleum and petrochemical plants. These standards require process changes, emission control devices, monitoring, and other activities designed to reduce emissions from about 200 benzene sources in the United States. Not all sources are covered, since the EPA approach has been to concentrate on the main sources that account for the great majority of emissions. The EPA estimated that these new regulations will reduce emissions to one-third of their present levels. Petitions to include other sources have been denied by the EPA.

11.5   BENZENE AND WATER POLLUTION

In addition to administering the Clean Air Act, as in the example above, the EPA is also responsible for carrying out the provisions of the Safe Drinking Water Act (SDWA). This legislation requires the EPA to set standards for maximum levels of toxic chemicals in drinking water. However, after an initial flurry of activity in the mid-1970s resulting in standards for about 20 chemicals, there was a long lull in standard setting. In the past few years, however, activity has increased considerably in this area, and regulatory actions have been initiated with respect to about 50 more chemicals. In 1986, Congress passed amendments to the Safe Drinking Water Act that require the EPA to establish standards for 83 chemicals within the next three years.

Benzene is in the group of 50, and is among a group of 8 for which maximum contaminant levels have recently been adopted. The regulation, which takes effect in January 1989, sets a maximum contaminant level of 0.005 ppm or 5 ppb. This level is not based on the results of a quantitative risk calculation. Instead, it is the lowest level that can be consistently measured analytically. In addition to requiring the new standards, the revised SDWA mandates monitoring of drinking water supplies for a much larger number of chemicals than are presently measured.

Although there has been a greatly heightened awareness of the toxic effects of benzene in recent years and much regulatory activity, management of benzene has not become much more stringent than it was about 10 years ago. It does appear that this situation is likely to change in the near future. However, it is also clear that it is very unlikely this chemical will be banned and that benzene will share the fate that cyclamates suffered and that asbestos may eventually face.

This fourth case study provides the last example of risk assessment and risk management and how they are related. Although the precise details could not be provided in a book of this type, it is hoped that enough information has been provided to make it possible to understand the similarities and differences among these cases. These similarities and differences will provide a vehicle for the discussion in the last chapter of the way that principles of toxicology and societal attitudes and values interact.

12

Summary

By now it should be very clear that there is a large gap between our present levels of understanding in toxicology and the breadth of knowledge needed to make confident judgments about chemicals in our environment. The science of toxicology is a biological science and, as such, has to deal with living organisms that are continually changing. In addition, toxicologists must work with a dynamic system where even the individual parts have not been totally characterized. It is as if engineers were trying to understand a mechanical device that was in continual motion and where the individual parts were not clearly delineated. Actually, it is even worse, since toxicologists are asked to make judgments about organisms that they cannot study in a controlled situation—humans. It is as if these same engineers were further handicapped by having to look at a surrogate machine that shared some but not all of the attributes of the machine they wished to understand. Thus it is not too surprising that toxicology can only provide tentative answers to a wide variety of critical questions.

However, it must also be recognized that total understanding is not needed for knowledge to be gained and action to be taken. For example, aspirin was used for decades without any understanding of how it worked. It was clear, however, that there was a cause-effect relationship between aspirin administration and analgesic effects. Similarly, it has been possible to determine that certain chemicals cause specific effects in animals and humans even if the mechanisms behind the effects are not known. However, it is important to realize that establishing that aspirin had analgesic effects did not imply

101

that the magnitude of the effect was the same in all people or that all analgesic effects observed after aspirin administration were due to the intake of this drug. In fact, it is clear that some of its impact was due to the placebo effect, i.e., psychosomatic changes related to a belief in the efficacy of the agent. Similarly, even though a relationship between a chemical and a toxic effect is established, this does not mean that this will occur in every person or to the same extent in every person. Additionally, some individuals will show symptoms as a result of a reverse placebo effect, i.e., psychosomatic illness related to belief in the toxicity of the agent.

Thus a lack of understanding of the mechanism of action leads to a degree of uncertainty that cannot be avoided. This suggests that no matter how many tests are made or how many chemicals are tested, residual uncertainty will still remain. Since conclusions bearing some degree of uncertainty are the best that can be achieved under current conditions, it is clear that even greater uncertainty will be the rule in most situations. Some of the sources of this additional uncertainty have been dealt with in the chapter on risk assessment.

Partly as a result of this uncertainty in assessing risk, there is a significant amount of leeway in the action that society can take to deal with potentially toxic chemicals. At one extreme, it could be said that the amount of uncertainty is so great that all chemicals showing a certain degree of toxicity should be banned. This is obviously not possible, since many are naturally occurring, and is also not feasible because many of these compounds are essential to our way of life. At the other extreme, it could be said that none need to be regulated; instead, the evidence that is available could be made public and everyone could make their own choices. At first glance, this appears to provide the greatest individual freedom of choice but, on reflection, only provides choice to those who have the education and degree of affluence to make such choices meaningful. Thus risk management must take a middle course.

The middle course that has been employed has emphasized caution and has leaned toward some regulation of all chemicals which exhibit a significant degree of toxicity. Thus conservative assumptions have been used in assessing possible toxicity from both acute and chronic exposures, i.e., a worst possible case or at least a very bad possible case has been utilized as the basis for actions that are taken. However, this conservatism in the risk assessment has sometimes been counterbalanced by other considerations which are written into regulations dealing with various types of exposure situations. For example, economic and technical feasibility are significant criteria in some legislation which is enforced by the EPA and OSHA. The social good might be considered a major influence in other management decisions, such as cigarette warnings. In addition, the public perception can have a crucial influence on the amount and severity of regulation to which a given chemical is subject.

The four examples that were discussed reflect not only the factors that enter into risk management decisions of various types but also something of the evolution of both risk assessment and risk management during the past two decades. The decision to ban cyclamates was made at a time when present criteria for establishing carcinogenicity were not in place. It is doubtful that the studies presented to support the ban would carry similar weight today. In addition, the FDA has been moving away from a strict interpretation of the Delaney clause and trying to establish the principle that very small, or insignificant, risks need not be regulated. The present case in point is the use of methylene chloride in coffee decaffeination. The evidence about its carcinogenicity is equivocal, but the FDA argument is that even if one accepts the positive studies, they do not indicate a significant risk, e.g., one in a million or greater.

The cyclamate issue does, however, reflect a national preoccupation with cancer that persists to the present. The current public attitude appears to favor strict regulation of small cancer risks in preference to larger risks of other types of toxic effects, including reproductive ones. The cyclamate case also illustrates the importance of exposure. Since this material was used in food, it was potentially in everyone's diet and thus exposure was broad. The distinction between the treatment of cyclamates and saccharin illustrates that this criterion of exposure is not paramount, since saccharin is probably much more widely used now than cyclamates ever were.

Perhaps, the cyclamate-saccharin distinction illustrates a different, but equally important consideration: the perceived desirability of the chemical. The word desirability is used deliberately, since it is not necessarily a real benefit that is involved. In this case, the attraction of the sugar substitutes was the desired sweetness without the undesirable calories. There was no good evidence, however, which showed that intake of foods with artifical sweeteners led to weight loss or even overall decreases in caloric intake. Thus the purported benefit, reduced intake of calories, was probably not real. If this analysis is correct, the reason for the ban on cyclamate but not saccharin could be the existence of an alternative at the time of the cyclamate ban but not when saccharin was under investigation.

Let us examine the second example, asbestos, in light of these considerations. Initially, asbestos was found to be linked with lung disease in the work environment. The effect was not cancer and was not inconsistent with effects of substances used in other industries at that time. Thus measures were taken to reduce exposure, but not to minimize them to the fullest extent feasible. Usage of the material increased with time and exposure spread. However, the most severe effects appeared in those who were exposed heavily in their jobs, and so other types of exposures were ignored. After the links with lung cancer and mesothelioma were demonstrated, there was more stringent regulation. However, this was again limited to the workplace. Only after there was a concern that limited exposure in the general population could lead to the carcinogenic effects was action initiated to limit environmental exposure.

In view of the uncertainty as to the minimum amount of exposure that could produce cancer, even very low levels of asbestos became suspect. Perhaps more critically, the revelation that asbestos was common in schools gave the situation new gravity. Here was the specter of very young children exposed to a substance which might have dire adverse effects decades from now. Action has not been that swift, however, and the desirability of the products made with asbestos has helped to temper the regulatory process. It is too early to tell if the proposed phaseout of asbestos will occur as planned.

Formaldehyde, the third chemical discussed, illustrates a somewhat different set of circumstances. The main concern has been about indoor exposures, but in the home and not in the workplace. It has also been shown to be a carcinogen in rodents, but the studies have been carefully examined and the significance minimized for the general population. As a result, little general action has been taken, although some proposals for limiting workplace exposure have been made.

It is hard to determine why risk management actions have been so limited. Again, this is a rather ubiquitous chemical, so that any type of ban would have a tremendous impact on a wide variety of common products. Another factor might be the natural occurrence of formaldehyde in the human body. However, public uneasiness about this compound is quite evident from questions that are often raised at public meetings and resistance to siting of formaldehyde-producing facilities in any community. Perhaps, stronger actions will be taken in the future.

The fourth example, benzene, again illustrates the way that management of substances that are considered solely workplace hazards often occurs. In this case, as in the last, gradual steps were taken to curtail worker exposure even after suggestions were made that benzene is a carcinogen. Even now, when benzene is accepted as a human carcinogen, no ban is proposed. Instead, a lowering of levels has been suggested by a government agency at the legal urging of outside groups. Benzene is a substance that is widely used in manufacture and is also found in our food and in our fuel. In this case, there has been no dramatic instance of exposure to benzene such as the schools provide with respect to asbestos. Benzene-containing products, such as gasoline, are desirable. Interestingly, benzene is produced in cigarette smoke, so that a total ban of this material would undoubtedly also necessitate stricter regulations on tobacco products.

Together, these examples amply reveal the complexity of managing toxic substances in our modern society. They also illustrate the even more difficult problems facing citizens who have to make choices on an individual basis. For example, there is the question of whether or not to move when a source of contamination is found somewhere in the vicinity. For another, there is the question of whether or not to accept employment in a situation where some exposure to a toxic chemical is possible. These dilemmas cannot be solved in a general way. It is no good to ask a toxicologist or public health official what they would or would not do in the same situation.

Individual choice in these kinds of situations requires consideration of many of the same factors as discussed in societal management situations. How much will it cost to move? What are risks of moving to another location which may be less well investigated than the one from which the move is being made? Are there children involved? However, beyond this there are even greater difficulties which result from our lack of toxicological knowledge. It is one thing to talk about risks of one in a million in the abstract, but what if someone in one's immediate family is the one? It is also one thing to talk about testing one substance at a time, but what about the real exposure to many chemicals that may have occurred in your case.

These questions are unanswerable, as are many of the questions that we face in life. However, it is easier for citizens to come to a conclusion that they can feel comfortable about if the uncertainties are recognized up front and no unrealistic expectations about what can be known are held. An understanding of toxicology provides some of this support and allows individuals to distinguish among various situations which involve varying degrees of uncertainty. In addition, a grasp of how the standards and other values that govern toxic chemicals are established provides a better idea of how to use these yardsticks. Thus knowledge of both the risk assessment and risk management aspects of environmental toxicology are essential for individual choice as well as public policy.

It is hoped that this small volume has provided an understanding of our current understanding of toxicology and how this knowledge is used by our society. A greater awareness of these aspects of toxicology is essential in a society in which toxic chemicals have such a heavy impact. Since everyone shares the same environment, each citizen's individual choices and votes have an effect on everyone else. Knowledge can contribute to the elimination of adversarial interactions and the beginning of a more enlightened and cooperative era in dealing with toxic chemicals.

Appendix A

National Primary Drinking Water Standards

Chemical contaminants:	Maximum contaminant level (micrograms/liter or parts per billion)
Arsenic	50
Barium	1,000
Cadmium	10
Chromium	50
Lead	50
Mercury	2
Nitrate (as Nitrogen)	10,000
Selenium	10
Silver	50
Fluoride	4,000
Endrin	0.2
Lindane	4
Methoxychlor	100
Toxaphene	5
2,4-D	100
2,4,5-TP	10
Trihalomethanes	100

Other contaminants	Maximum allowable concentration
Coliform bacteria	1 per 100 milliliters (mean)
Radionuclides	
Radium 226 and 228 (combined)	5 picoCuries/liter
Gross alpha particle activity	15 picoCuries/liter
Gross beta particle activity	4 millirem per year
Organic chemicals turbidity	1 up to 5 turbidity units

Source: Code of Federal Regulations, Title 40, Part 141.

Appendix B

National Ambient Air Quality Standards

	Primary standard	Secondary standard
Particulate matter		
annual arithmetic mean	50 μg/m³	50 μg/m³
maximum 24-hour concentration[a]	150 μg/m³	150 μg/m³
Sulfur dioxide		
annual arithmetic mean	80 μg/m³ (0.03 ppm)	
maximum 24-hour concentration[a]	365 μg/m³ (0.14 ppm)	
maximum 3-hour concentration		1300 μg/m³ (0.50 ppm)
Carbon monoxide		
maximum 8-hour concentration[a]	10 mg/m³ (9 ppm)	10 mg/m³ (9 ppm)
maximum 1-hour concentration[a]	40 mg/m³ (35 ppm)	40 mg/m³ (35 ppm)
Ozone		
maximum daily 1-hour concentration[a]	235 μg/m³ (0.12 ppm)	235 μg/m³ (0.12 ppm)
Nitrogen dioxide		
annual arithmetic mean	100 μg/m³ (0.05 ppm)	100 μg/m³ (0.05 ppm)
Lead		
maximum calendar quarter average[a]	1.5 μg/m³	1.5 μg/m³
Hydrocarbons[b]		
maximum 3-hour concentration (6–9 a.m.)	160 μg/m³ (0.24 ppm)	160 μg/m³ (0.24 ppm)

[a]Not to be exceeded more than once a year per site.
[b]A non-health standard used as a guide for ozone control.

Note: ppm = parts per million.

Source: Code of Federal Regulations, Title 40, Part 50.

National Emission Standards for Hazardous Air Pollutants

Asbestos

asbestos mills or asbestos manufacturing	no visible emissions to the outside air or use of specified air cleaning procedures

Benzene

petroleum refineries or chemical manufacturing plants	Monthly monitoring of valves and pumps; installation of leak-prevention equipment; repair of leaks within 15 days

Beryllium*

maximum 24-hour emission or 30-day average in the vicinity of the source	10 grams 0.01 $\mu g/m^3$

Mercury

maximum 24-hour emission from ore processing or mercury cell chlor-alkali	2300 g
maximum 24-hour emission from sludge incineration, sludge drying, or combination that processes wastewater treatment sludges	3200 g

Vinyl chloride

ethylene chloride purification waste gas maximum	10 ppm
oxychlorination reactor	0.2 g/kg of the 100% ethylene dichloride product
vinyl chloride plant maximum	10 ppm

Source: Code of Federal Regulations, Title 40, Part 61.
*There is a separate standard for rocket motor firing.

Appendix C

National Occupational Standards

Chemical Name	Permissible Exposure Limit (8-hr TWA)*	Target Organ
Acetaldehyde	200 ppm (360 mg/m³)	Respiratory system, skin, kidneys
Acetic Acid	10 ppm (25 mg/m³)	Respiratory system, skin, eyes, teeth
Acetic anhydride	5 ppm (20 mg/m³)	Respiratory system, eyes, skin
Acetone	1000 ppm (2400 mg/m³)	Respiratory system, skin
Acetonitrile	40 ppm (70 mg/m³)	Kidneys, liver, CVS, CNS, lungs, skin, eyes
Acetylene tetrabromide	1 ppm (14 mg/m³)	Eyes, upper respiratory system, liver
Acrolein	0.1 ppm (0.25 mg/m³)	Heart, eyes, skin, respiratory system
Acrylamide	0.3 mg/m³	CNS, PNS, skin, eyes
Acrylonitrile	2 ppm; 10 ppm ceiling, 15 min	CVS, liver, kidneys, CNS, skin, brain tumor, lung and bowel cancer
Aldrin	0.25 mg/m³	Cancer, CNS, liver, kidneys, skin
Allyl alcohol	2 ppm (5 mg/m³)	Eyes, skin, respiratory system
Allyl chloride	1 ppm (3 mg/m³)	Respiratory systems, skin, eyes, liver, kidneys
Allyl glycidyl ether	10 ppm/ceiling (45 mg/m³)	Respiratory system, skin
2-Aminopyridine	0.5 ppm (2 mg/m³)	CNS, respiratory sytem
Ammonia	50 ppm (35 mg/m³)	Respiratory system, eyes
Ammonium sulfamate	15 mg/m³	None known

Source: Adapted from NIOSH Pocket Guide to Chemical Hazards, September 1985. CNS = central nervous system; PNS = peripheral nervous system; CVS = cardiovascular system; GI = gastrointestinal; TWA = time-weighted average.

113

National Occupational Standards   Continued

Chemical Name	Permissible Exposure Limit (8-hr TWA)*	Target Organ
n-Amyl acetate	100 ppm (525 mg/m³)	Eyes, skin, respiratory system
sec-Amyl acetate	125 ppm (650 mg/m³)	Respiratory system, eyes, skin
Aniline	5 ppm (19 mg/m³)	Blood, CVS, liver, kidneys
Anisidine (o-,p-isomers)	0.5 mg/m³	Blood, kidneys, liver, CVS
Antimony and compounds (as Sb)	0.5 mg/m³	Respiratory sytem, CVS, skin, eyes
ANTU	0.3 mg/m³	Respiratory sytem
Arsenic and compounds (as As)	10 μg/m³	Liver, kidneys, skin, lungs, lymphatic system
Arsine	0.05 ppm (0.2 mg/m³)	Blood, kidneys, liver
Asbestos	0.2 fibers/cc	Lungs
Azinphos-methyl	0.2 mg/m³	Respiratory system, CNS, CVS, blood cholinesterase
Barium (soluble compounds as Ba)	0.5 mg/m³	Heart, CNS, skin, respiratory system, eyes
Benzene	10 ppm; 50 ppm ceiling (10 min)	Blood, CNS, skin, bone marrow, eyes, respiratory system
Benzoyl peroxide	5 mg/m³	Skin, respiratory system, eyes
Benzyl chloride	1 ppm (5 mg/m³)	Eyes, respiratory sytem, skin
Beryllium and compounds (as Be)	20 μg/m³; 5.0 μg/m³ ceiling; 25 μg/m³ (30-min ceiling)	Lung, skin, eyes, mucous membranes
Boron oxide	15 mg/m³	Skin, eyes
Boron trifluoride	1 ppm ceiling (3 mg/m³)	Respiratory system, kidneys, eyes, skin
Bromine	0.1 ppm (0.7 mg/m³)	Respiratory system, eyes, CNS
Bromoform	0.5 ppm (5 mg/m³)	Skin, liver, kidneys, respiratory system, CNS
Butadiene	1000 ppm (2200 mg/m³)	Eyes, respiratory sytem, CNS
2-Butanone	200 ppm (590 mg/m³)	CNS, lungs

National Occupational Standards   Continued

Chemical Name	Permissible Exposure Limit (8-hr TWA)*	Target Organ
2-Butoxy ethanol	50 ppm (240 mg/m³)	Liver, kidneys, lymphoid system, skin, blood, eyes, respiratory system
Butyl acetate	150 ppm (710 mg/m³)	Eyes, skin, respiratory system
sec-Butyl acetate	200 ppm (950 mg/m³)	Eyes, skin, respiratory system
tert-Butyl acetate	200 ppm (950 mg/m³)	Respiratory system, eyes, skin
Butyl alcohol	100 ppm (300 mg/m³)	Skin, eyes, respiratory system
sec-Butyl alcohol	150 ppm (450 mg/m³)	Eyes, skin, CNS
tert-Butyl alcohol	100 ppm (300 mg/m³)	Eyes, skin
Butylamine	5 ppm ceiling (15 mg/m³)	Respiratory system, skin, eyes
tert-Butyl chromate (as CrO3)	0.1 mg/m³ ceiling	Respiratory system, skin, eyes, CNS
n-Butyl glycidyl	50 ppm (270 mg/m³)	Eyes, skin, respiratory system, CNS
Butyl mercaptan	10 ppm (35 mg/m³)	Respiratory system; in animals: CNS, liver, kidneys
p-tert-Butyltoluene	10 ppm (60 mg/m³)	CVS, CNS, skin, bone marrow, eyes, upper respiratory system
Cadmium dust (as Cd)	0.2 mg/m³; 0.6 mg/m³ ceiling	Respiratory system, kidneys, prostate, blood
Cadmium fume (as Cd)	0.1 mg/m³; 0.3 mg/m³ ceiling	Respiratory system, kidneys, blood
Calcium arsenate	10 μg/m³	Eyes, respiratory system, liver, skin, lymphatics, CNS
Calcium oxide	5 mg/m³	Respiratory system, skin, eyes
Camphor	2 ppm (12 mg/m³)	CNS, eyes, skin, respiratory system
Carbaryl (Sevin)	5 mg/m³	Respiratory system, CNS, CVS, skin
Carbon black	3.5 mg/m³	None known
Carbon dioxide	5000 ppm	Lungs, skin, CVS

National Occupational Standards   Continued

Chemical Name	Permissible Exposure Limit (8-hr TWA)*	Target Organ
Carbon disulfide	20 ppm, 30 ppm ceiling, 100 ppm; 30-min ceiling	CNS, PNS, CVS, eyes, kidneys, skin
Carbon monoxide	50 ppm (55 mg/m³)	CVS, lungs, blood, CNS
Carbon tetrachloride	10 ppm; 25 ppm ceiling; 200 ppm; 5-min/4-hr peak	CNS, eyes, lungs, liver, kidneys, skin
Chlordane	0.5 mg/m³	CNS, eyes, lungs, liver, kidneys, skin
Chlorinated camphene	0.5 mg/m³	CNS, skin
Chlorinated diphenyl oxide	0.5 mg/m³	Skin, liver
Chlorine	1 ppm ceil (3 mg/m³)	Respiratory system
Chlorine dioxide	0.1 ppm (0.3 mg/m³)	Respiratory system, eyes
Chlorine trifluoride	0.1 ppm ceiling (0.4 mg/m³)	Skin, eyes
Chloro-acetaldehyde	1 ppm ceiling (3 mg/m³)	Eyes, skin, respiratory system
alpha-Chloro-acetophenone	0.05 ppm (0.3 mg/m³)	Eyes, skin, respiratory system
Chlorobenzene	75 ppm (350 mg/m³)	Respiratory system, eyes, skin, CNS, liver
o-Chloro-benzylidene malonitrile	0.05 ppm (0.4 mg/m³)	Respiratory system, skin, eyes
Chlorobromo-methane	200 ppm (1050 mg/m³)	Skin, liver, kidneys, respiratory system, CNS
Chlorodiphenyl (42% chlorine)	1 mg/m³	Skin, eyes, liver
Chlorodiphenyl (54% chlorine)	0.5 mg/m³	Skin, eyes, liver
Chloroform	50 ppm (240 mg/m³)	Liver, kidneys, heart, eyes, skin
1-Chloro-1-nitropropane	20 ppm (100 mg/m³)	In animals: respiratory system, liver, kidneys, CVS
Chloropicrin	0.1 ppm (0.7 mg/m³)	Respiratory system, skin, eyes
Chloroprene	25 ppm (90 mg/m³)	Respiratory system, skin, eyes
Chromic acid and chromates (as CrO₃)	0.1 mg/m³ ceiling	Blood, respiratory system, liver, kidneys, eyes, skin

National Occupational Standards   Continued

Chemical Name	Permissible Exposure Limit (8-hr TWA)*	Target Organ
Chromium, metal and insoluble salts (as Cr)	1 mg/m³	Respiratory system
Chromium, soluble chromic, chromous salts (as Cr)	0.5 mg/m³ (NIOSH)	Skin
Coal tar pitch volatiles	0.2 mg/m³ (benzene soluble fraction)	Respiratory system, bladder, kidneys, skin
Cobalt metal, fume, and dust (as Co)	0.1 mg/m³	Respiratory system, skin
Copper dust and mist (as Cu)	1 mg/m³	Respiratory system, skin, liver, increased risk with Wilson's disease, kidneys
Copper fume (as Cu)	0.1 mg/m³	Respiratory system, skin, eyes, increased risk with Wilson's disease
Crag herbicide	15 mg/m³	None known
Cresol	5 ppm (22 mg/m³)	CNS, respiratory system, liver, kidneys, skin, eyes
Crotonaldehyde	2 ppm (6 mg/m³)	Respiratory system, eyes, skin
Cumene	50 ppm (245 mg/m³)	Eyes, upper respiratory system, skin, CNS
Cyanides (as CN)	5 mg/m³	CVS, CNS, liver, kidneys, skin
Cyclohexane	300 ppm (1050 mg/m³)	Eyes, respiratory system, skin, CNS
Cyclohexanol	500 ppm (200 mg/m³)	Eyes, respiratory system, skin
Cyclohexanone	50 ppm (200 mg/m³)	Respiratory system, eyes, skin, CNS
Cyclohexene	300 ppm (1015 mg/m³)	Skin, eyes, respiratory system
Cyclopentadiene	75 ppm (200 mg/m³)	Eyes, respiratory system
2,4-D	10 mg/m³	Skin, CNS
DDT	1 mg/m³	CNS, kidneys, liver, skin, PNS
Decaborane	0.05 ppm (0.3 mg/m³)	CNS
Demeton	0.1 mg/m³	Respiratory system, CVS, CNS, skin, eyes, blood cholinesterase

National Occupational Standards   Continued

Chemical Name	Permissible Exposure Limit (8-hr TWA)*	Target Organ
Diacetone alcohol	50 ppm (240 mg/m³)	Eyes, skin, respiratory system
Diazomethane	0.2 ppm (0.4 mg/m³)	Respiratory system, eyes, skin
Diborane	0.1 ppm (0.1 mg/m³)	Respiratory system, CNS
Dibromo-chloropropane	1 ppb	CNS, skin, liver, kidney, spleen, reproductive system, digestive system
Dibutyl phosphate	1 ppm (5 mg/m³)	Respiratory system, skin
Dibutylphthalate	5 mg/m³	Respiratory sytem, GI tract
o-Dichloro-benzene	50 ppm ceiling (300 mg/m³)	Liver, kidneys, skin, eyes
p-Dichlorobenzene	75 ppm (450 mg/m³)	Liver, respiratory system, eyes, kidneys, skin
Dichlorodifluoro-methane	1000 ppm (4950 mg/m³)	CVS, PNS
1,3-Dichloro-5,5-dimethyl hydantoin	0.2 mg/m³	Respiratory system, eyes
1,1-Dichloroethane	100 ppm (400 mg/m³)	Skin, liver, kidneys
1,2-Dichloro-ethylene	200 ppm (790 mg/m³)	Respiratory system, eyes, CNS
Dichloroethyl ether	15 ppm ceiling (90 mg/m³)	Respiratory system, skin, eyes
Dichloromono-fluoromethane	1000 ppm (4200 mg/m³)	Respiratory system, CVS
1,1- Dichloro-1-nitroethane	10 ppm ceiling (60 mg/m³)	Lungs
Dichlorotetra-fluoroethane	1000 ppm (7000 mg/m³)	Respiratory system, CVS
Dichlorvos	1 mg/m³	Respiratory system, CVS, CNS, eyes, skin, blood cholinesterase
Dieldrin	0.25 mg/m³	CNS, liver, kidneys, skin
Diethylamine	25 ppm (75 mg/m³)	Respiratory system, skin, eyes
Diethylamino-ethanol	10 ppm (50 mg/m³)	Respiratory system, skin, eyes
Difluorodibromo-methane	100 ppm (860 mg/m³)	Skin, respiratory system
Diglycidyl ether	0.5 ppm (2.8 mg/m³)	Skin, eyes, respiratory system

National Occupational Standards   Continued

Chemical Name	Permissible Exposure Limit (8-hr TWA)*	Target Organ
Diisobutyl ketone	50 ppm (290 mg/m³)	Respiratory system, skin, eyes
Diisopropylamine	5 ppm (20 mg/m³)	Respiratory system, skin, eyes
Dimethyl acetamide	10 ppm (35 mg/m³)	Liver, skin
Dimethylamine	10 ppm (18 mg/m³)	Respiratory system, skin, eyes
Dimethylaniline	5 ppm (25 mg/m³)	Blood, kidneys, liver, CVS
Dimethyl-1,2-dibromo-2,2-dichlorethyl phosphate	3 mg/m³	Respiratory system, CNS, CVS, skin, eyes, blood cholinesterase
Dimethyl formamide	10 ppm (30 mg/m³)	Liver, kidneys, CVS, skin
1,1-Dimethyl-hydrazine	0.5 ppm (1 mg/m³)	CNS, liver, GI tract, blood, respiratory system, eyes, skin
Dimethylphthalate	5 mg/m³	Respiratory system, GI tract
Dimethylsulfate	1 ppm (5 mg/m³)	Eyes, respiratory system, liver, kidneys, CNS, skin
Dinitrobenzene (all isomers)	1 mg/m³	Blood, liver, CVS, eyes, CNS
Dinitro-o-cresol	0.2 mg/m³	CVS, endocrine system, eyes
Dinitrotoluene	1.5 mg/m³	Blood, liver, CVS
Di-sec-octyl phthalate	5 mg/m³	Eyes, upper respiratory system, GI tract
Dioxane	100 ppm (360 mg/m³)	Liver, kidneys, skin, eyes
Diphenyl	0.2 ppm (1 mg/m³)	Liver, skin, CNS, upper respiratory system, eyes
Dipropylene glycol methyl ether	100 ppm (600 mg/m³)	Respiratory system, eyes
Endrin	0.1 mg/m³	CNS, liver
Epichlorohydrin	5 ppm (19 mg/m³)	Respiratory system, lungs, skin, kidneys
EPN	0.5 mg/m³	Respiratory system, CVS, CNS, eyes, skin, blood cholinesterase
Ethanolamine	3 ppm (6 mg/m³)	Skin, eyes, respiratory system

National Occupational Standards   Continued

Chemical Name	Permissible Exposure Limit (8-hr TWA)*	Target Organ
2-Ethoxyethanol	200 ppm (740 mg/m³)	In animals: lungs, eyes, blood, kidneys, liver
2-Ethoxyethyl-acetate	100 ppm (540 mg/m³)	Respiratory system, eyes, GI tract
Ethyl acetate	400 ppm (1400 mg/m³)	Eyes, skin, respiratory system
Ethylamine	10 ppm (18 mg/m³)	Respiratory system, eyes, skin
Ethyl acrylate	25 ppm (100 mg/m³)	Respiratory system, eyes, skin
Ethyl benzene	100 ppm (435 mg/m³)	Eyes, upper respiratory system, skin, CNS
Ethyl bromide	200 ppm (890 mg/m³)	Skin, liver, kidneys, respiratory system, CVS, CNS
Ethyl butyl ketone	50 ppm (230 mg/m³)	Eyes, skin, respiratory system
Ethyl chloride	1000 ppm (2600 mg/m³)	Liver, kidneys, respiratory system, CVS
Ethylene chloro-hydrin	5 ppm (16 mg/m³)	Respiratory system, liver, kidneys, CNS, skin, CVS
Ethylenediamine	10 ppm (25 mg/m³)	Respiratory system, liver, kidneys, skin
Ethylene dibromide	10 ppm; 30 ppm ceiling; 50 ppm, 5-min peak	Respiratory system, liver, kidneys, skin, eyes
Ethylene dichloride	50 ppm; 100 ppm ceiling; 200 ppm peak	Kidneys, liver, eyes, skin, CNS
Ethylene glycol dinitrate	1 mg/m³ ceiling	CVS, blood, skin
Ethylene oxide	1 ppm (1.8 mg/m³)	Eyes, blood, respiratory system, liver, CNS, kidneys
Ethyl ether	400 ppm (1200 mg/m³)	CNS, skin, respiratory system, eyes
Ethyl formate	100 ppm (300 mg/m³)	Eyes, respiratory system
Ethyl mercaptan	10 ppm ceiling (25 mg/m³)	Respiratory system; in animals: liver, kidneys
n-Ethylmorpholine	20 ppm (94 mg/m³)	Respiratory system, eyes, skin
Ethyl silicate	100 ppm (850 mg/m³)	Respiratory system, liver, kidneys, blood, skin
Ferbam	15 mg/m³	Respiratory system, skin, GI tract

National Occupational Standards   Continued

Chemical Name	Permissible Exposure Limit (8-hr TWA)*	Target Organ
Ferrovanadium dust	1 mg/m³	Respiratory system, eyes
Fluorides (as F)	2.5 mg/m³	Eyes, respiratory system, CNS, skeleton, kidneys, skin
Fluorine	0.1 ppm (0.2 mg/m³)	Respiratory system, eyes, skin; in animals: liver, kidneys
Fluorotrichloro-methane	1000 ppm (5600 mg/m³)	CVS, skin
Formaldehyde	3 ppm; 5 ppm ceiling; 10 ppm 30-min ceiling	Respiratory system, eyes, skin
Formic acid	5 ppm (9 mg/m³)	Respiratory system, skin, kidneys, liver, eyes
Furfural	5 ppm (20 mg/m³)	Eyes, respiratory system, skin
Furfuryl alcohol	50 ppm (200 mg/m³)	Respiratory system
Glycidol	50 ppm (150 mg/m³)	Eyes, skin, respiratory system, CNS
Graphite (natural)	15 mppcf	Respiratory system, CVS
Hafnium and compounds (as Hf)	0.5 mg/m³	Eyes, skin, mucous membranes
Heptachlor	0.5 mg/m³	In animals: CNS, liver
Heptane	500 ppm (2000 mg/m³)	Skin, respiratory system, PNS
Hexachloroethane	1 ppm (10 mg/m³)	Eyes
Hexachloro-naphthalene	0.2 mg/m³	Liver, skin
Hexane	500 ppm (1800 mg/m³)	Skin, eyes, respiratory system, lungs
2-Hexanone	100 ppm (410 mg/m³)	CNS, skin, respiratory system
Hexone	100 ppm (410 mg/m³)	Respiratory system, eyes, skin, CNS
sec-Hexyl acetate	50 ppm (300 mg/m³)	CNS, eyes
Hydrazine	1 ppm (1.3 mg/m³)	CNS, respiratory system, skin, eyes
Hydrogen bromide	3 ppm (10 mg/m³)	Respiratory system, eyes, skin
Hydrogen chloride	5 ppm ceiling (7 mg/m³)	Respiratory system, skin, eyes

National Occupational Standards   Continued

Chemical Name	Permissible Exposure Limit (8-hr TWA)*	Target Organ
Hydrogen cyanide (as CN)	5 mg/m³	CNS, CVS, liver, kidneys
Hydrogen fluoride	3 ppm (2 mg/m³)	Eyes, respiratory system, skin
Hydrogen peroxide	1 ppm (1.4 mg/m³)	Eyes, skin, respiratory system
Hydrogen selenide	0.05 ppm (0.2 mg/m³)	Respiratory system, eyes
Hydrogen sulfide	20 ppm ceiling; 50 ppm; 10-min peak	Respiratory system, eyes
Hydroquinone	2 mg/m³	Eyes, respiratory system, skin, CNS
Iodine	0.1 ppm ceiling (1 mg/m³)	Respiratory system, eyes, skin, CNS, CVS
Iron oxide fume	10 mg/m³	Respiratory system
Isoamyl acetate	100 ppm (525 mg/m³)	Eyes, skin, respiratory system
Isoamyl alcohol	100 ppm (360 mg/m³)	Eyes, skin, respiratory system
Isobutyl acetate	150 ppm (700 mg/m³)	Skin, eyes, respiratory system
Isobutyl alcohol	100 ppm (300 mg/m³)	Eyes, skin, respiratory system
Isophorone	25 ppm (140 mg/m³)	Respiratory system
Isopropyl acetate	250 ppm (950 mg/m³)	Eyes, skin, respiratory system
Isopropyl alcohol	400 ppm (980 mg/m³)	Eyes, skin, respiratory system
Isopropylamine	5 ppm (12 mg/m³)	Respiratory system, skin, eyes
Isopropyl ether	500 ppm (2100 mg/m³)	Respiratory system, skin
Isopropyl glycidyl ether	50 ppm (240 mg/m³)	Eyes, skin, respiratory system
Ketene	0.5 ppm (0.9 mg/m³)	Respiratory system, eyes, skin
Lead, inorganic fumes and dusts (as Pb)	0.05 mg/m³	GI tract, CNS, kidneys, blood, gingival tissue
Lead arsenate	0.05 mg/m³ (as lead)	GI tract, CNS, kidneys, blood, gingival tissue, lymphatics, skin
Lindane	0.5 mg/m³	Eyes, CNS, blood, liver, kidneys, skin

National Occupational Standards   Continued

Chemical Name	Permissible Exposure Limit (8-hr TWA)*	Target Organ
Lithium hydride	0.025 mg/m³	Respiratory system, skin, eyes
LPG	1000 ppm (1800 mg/m³)	Respiratory system, CNS
Magnesium oxide fume	15 mg/m³	Respiratory system, eyes
Malathion	15 mg/m³	Respiratory system, liver, blood cholinesterase, CNS, CVS, GI
Maleic anhydride	0.25 ppm (1 mg/m³)	Eyes, respiratory system, skin
Manganese and compounds (as Mn)	5 mg/m³ceiling	Respiratory system, CNS, blood, kidneys
Mercury and inorganic compounds (as Hg)	0.1 mg/m³ceiling	Skin, respiratory system, CNS, kidneys, eyes
Mercury, (organo) alkyl compounds (as Hg)	0.01 mg/m³; 0.04 mg/m³ ceiling	CNS, kidneys, eyes, skin
Mesityl oxide	25 ppm (100 mg/m³)	Eyes, skin, respiratory system, CNS
Methoxychlor	15 mg/m³	None known
Methyl acetate	200 ppm (610 mg/m³)	Respiratory system, skin, eyes
Methyl acetylene	1000 ppm (1650 mg/m³)	CNS
Methyl acetylene-propadiene mixture	1000 ppm (1800 mg/m³)	CNS, skin, eyes
Methyl acrylate	10 ppm (35 mg/m³)	Respiratory system, eyes, skin
Methylal	1000 ppm (3100 mg/m³)	Skin, respiratory system, CNS
Methyl alcohol	200 ppm (260 mg/m³)	Eyes, skin, CNS, GI tract
Methylamine	10 ppm (12 mg/m³)	Respiratory system, eyes, skin
Methyl (n-amyl) ketone	100 ppm (465 mg/m³)	Eyes, skin, respiratory system, CNS, PNS
Methyl bromide	20 ppm (80 mg/m³)	CNS, respiratory system, skin, eyes
Methyl cellosolve	25 ppm (80 mg/m³)	CNS, blood, skin, eyes, kidneys

National Occupational Standards   Continued

Chemical Name	Permissible Exposure Limit (8-hr TWA)*	Target Organ
Methyl cellosolve acetate	25 ppm (120 mg/m³)	Kidneys, brain, CNS, PNS
Methyl chloride	100 ppm; 200 ppm ceiling; 300 ppm 5 min/3-hr peak	CNS, liver, kidneys, skin
Methyl chloroform	350 ppm (1900 mg/m³)	Skin, CNS, CVS, eyes
Methylcyclo-hexane	500 ppm (2000 mg/m³)	Respiratory system, skin
Methylcyclo-hexanol	100 ppm (470 mg/m³)	Respiratory system, skin, eyes; in animals: CNS, liver, kidneys
o-Methylcyclo-hexanone	100 ppm (460 mg/m³)	In animals: lungs, liver, kidneys, skin
Methylene bisphenyl isocyanate	0.02 ppm ceiling (0.2 mg/m³)	Respiratory system, eyes
Methylene chloride	500 ppm;1000 ppm ceiling 2000 ppm 5 min/2-hr peak	Skin, CVS, eyes, CNS
Methyl formate	100 ppm (250 mg/m³)	Eyes, respiratory system, CNS
5-Methyl-3-heptanone	25 ppm (130 mg/m³)	Eyes, skin, respiratory system, CNS
Methyl iodide	5 ppm (28 mg/m³)	CNS, skin, eyes
Methyl isobutyl carbinol	25 ppm (100 mg/m³)	Eyes, skin
Methyl isocyanate	0.02 ppm (0.05 mg/m³)	Respiratory system, eyes, skin
Methyl mercaptan	10 ppm 15-min ceiling (20 mg/m³)	Respiratory system, CNS
Methyl methacrylate	100 ppm (410 mg/m³)	Eyes, upper respiratory system, skin
alpha-Methyl styrene	100 ppm ceiling (480 mg/m³)	Eyes, respiratory system, skin
Mica (less than 1% quartz)	20 mppcf	Lungs
Molybdenum soluble compounds (as Mo)	5 mg/m³	Respiratory system; in animals: kidneys, blood
Molybdenum insoluble compounds (as Mo)	15 mg/m³	None known

National Occupational Standards Continued

Chemical Name	Permissible Exposure Limit (8-hr TWA)*	Target Organ
Monomethyl aniline	2 ppm (9 mg/m³)	Respiratory system, liver, kidneys, blood
Monomethyl hydrazine	0.2 ppm ceiling (0.35 mg/m³)	CNS, respiratory system, liver, blood, CVS, eyes
Morpholine	20 ppm (70 mg/m³)	Respiratory system, eyes, skin
Naphtha (coal tar)	100 ppm (400 mg/m³)	Respiratory system, eyes, skin
Naphthalene	10 ppm (50 mg/m³)	Eyes, blood, liver, kidneys, skin, RBC, CNS
Nickel, metal and soluble compounds (as Ni)	1 mg/m³	Nasal cavities, lungs, skin
Nickel carbonyl	0.001 ppm (0.007 mg/m³)	Lungs, paranasal sinus, CNS
Nicotine	0.5 mg/m³	CNS, CVS, lungs, GI tract
Nitric acid	2 ppm (5 mg/m³)	Eyes, respiratory system, skin, teeth
Nitric oxide	25 ppm (30 mg/m³)	Respiratory system
p-Nitroaniline	1 ppm (6 mg/m³)	Blood, heart, lungs, liver
Nitrobenzene	1 ppm (5 mg/m³)	Blood, liver, kidneys, CVS, skin
p-Nitrochloro-benzene	1 mg/m³	Blood, liver, kidneys, CVS
Nitroethane	100 ppm (310 mg/m³)	Skin
Nitrogen dioxide	5 ppm ceiling (9 mg/m³)	Respiratory system, CVS
Nitrogen trifluoride	10 ppm (29 mg/m³)	In animals: blood
Nitromethane	100 ppm (250 mg/m³)	Skin
1-Nitropropane	25 ppm (90 mg/m³)	Eyes, CNS
2-Nitropropane	25 ppm (90 mg/m³)	Respiratory system, CNS
Nitrotoluene	5 ppm (30 mg/m³)	Blood, CNS, CVS, skin, GI tract
Octachloro-naphthalene	0.1 mg/m³	Skin, liver
Octane	500 ppm (2350 mg/m³)	Skin, eyes, respiratory system
Oil mist (mineral)	5 mg/m³	Respiratory system, skin
Osmium tetroxide	0.002 mg/m³	Eyes, respiratory system, skin
Oxalic acid	1 mg/m³	Respiratory system, skin, kidneys, eyes

National Occupational Standards   Continued

Chemical Name	Permissible Exposure Limit (8-hr TWA)*	Target Organ
Oxygen difluoride	0.05 ppm (0.1 mg/m³)	Lungs, eyes
Ozone	0.1 ppm (0.2 mg/m³)	Eyes, respiratory system
Paraquat compounds	0.5 mg/m³	Eyes, respiratory system, heart, liver, kidneys, GI tract
Parathion	0.1 mg/m³	Respiratory system, CNS, CVS, eyes, skin, blood cholinesterase
Pentaborane	0.005 ppm (0.01 mg/m³)	CNS, eyes, skin
Pentachloro-naphthalene	0.5 mg/m³	Skin, liver, CNS
Pentachloro-phenol	0.5 mg/m³	CVS, respiratory system, eyes, liver, kidneys, skin, CNS
Pentane	1000 ppm (2950 mg/m³)	Skin, eyes, respiratory system
2-Pentanone	200 ppm (700 mg/m³)	Respiratory system, eyes, skin, CNS
Perchloromethyl mercaptan	0.1 ppm (0.8 mg/m³)	Eyes, respiratory system, liver, kidneys, skin
Perchloryl fluoride	3 ppm (13.5 mg/m³)	Respiratory system, skin, blood
Petroleum distillates (naphtha)	500 ppm (2000 mg/m³)	Skin, eyes, respiratory system, CNS
Phenol	5 ppm (19 mg/m³)	Liver, kidneys, skin
p-Phenylene diamine	0.1 mg/m³	Respiratory system, skin
Phenyl ether	1 ppm (7 mg/m³)	Eyes, skin, respiratory system
Phenyl ether-biphenyl mixture	1 ppm (7 mg/m³)	Eyes, skin, respiratory system
Phenyl glycidyl ether	10 ppm (60 mg/m³)	Skin, eyes, CNS
Phenylhydrazine	5 ppm (22 mg/m³)	Blood, respiratory system, liver, kidneys, skin
Phosdrin	0.1 mg/m³	Respiratory system, CNS, CVS, skin, blood cholinesterase
Phosgene	0.1 ppm (0.4 mg/m³)	Respiratory system, skin, eyes
Phosphine	0.3 ppm (0.4 mg/m³)	Respiratory system
Phosphoric acid	1 mg/m³	Respiratory system, eyes, skin

National Occupational Standards   Continued

Chemical Name	Permissible Exposure Limit (8-hr TWA)*	Target Organ
Phosphorus (yellow)	0.1 mg/m³	Respiratory system, liver, kidneys, jaw, teeth, blood, eyes, skin
Phosphorus pentachloride	1 mg/m³	Respiratory system, eyes, skin
Phosphorus pentasulfide	1 mg/m³	Respiratory system, CNS, eyes, skin
Phosphorus trichloride	0.5 ppm (3 mg/m³)	Respiratory system, eyes, skin
Phthalic anhydride	2 ppm (12 mg/m³)	Respiratory system, eyes, skin, liver, kidneys
Picric acid	0.1 mg/m³	Kidneys, liver, blood, skin, eyes
Pival	0.1 mg/m³	Blood prothrombin
Platinum (soluble salts as Pt)	0.002 mg/m³	Respiratory system, skin, eyes
Portland cement (less than 1% quartz)	50 mppcf	Respiratory system, eyes, skin
Propane	1000 ppm (1800 mg/m³)	CNS
n-Propyl acetate	200 ppm (840 mg/m³)	Respiratory system, eyes, skin, CNS
Propyl alcohol	200 ppm (500 mg/m³)	Skin, eyes, respiratory system, GI tract
Propylene dichloride	75 ppm (350 mg/m³)	Skin, eyes, respiratory system, liver, kidneys
Propyleneimine	2 ppm (5 mg/m³)	Eyes, skin
Propylene oxide	100 ppm (240 mg/m³)	Eyes, skin, respiratory system
N-Propyl nitrate	25 ppm (110 mg/m³)	None known
Pyrethrum	5 mg/m³	Respiratory system, skin, CNS
Pyridine	5 ppm (15 mg/m³)	CNS, liver, kidneys, skin, GI tract
Quinone	0.1 ppm (0.4 mg/m³)	Eyes, skin
Rhodium, metal fume and dust (as Rh)	0.1 mg/m³	None known
Rhodium, soluble salts (as Rh)	0.001 mg/m³	Eyes

National Occupational Standards   Continued

Chemical Name	Permissible Exposure Limit (8-hr TWA)*	Target Organ
Ronnel	15 mg/m³	Skin, liver, kidneys, blood plasma
Rotenone	5 mg/m³	CNS, eyes, respiratory system
Selenium and compounds (as Se)	0.2 mg/m³	Upper respiratory system, eyes, skin, liver, kidneys, blood
Selenium hexa-fluoride (as Se)	0.05 ppm (0.4 mg/m³)	None known
Silica (amorphous)	20 mppcf	Respiratory system
Silica (crystalline)	10 mg/m³/%SiO_2 +2 (resp. quartz)	Respiratory system
Silver, metal, and soluble compounds (as Ag)	0.01 mg/m³	Nasal septum, skin, eyes
Soapstone	20 mppcf	Lungs, CVS
Sodium fluoro-acetate	0.5 mg/m³	CVS, lungs, kidneys, CNS
Sodium hydroxide	2 mg/m³	Eyes, respiratory system, skin
Stibine	0.1 ppm (0.5 mg/m³)	Blood, liver, kidneys, lungs
Stoddard solvent	500 ppm (2900 mg/m³)	Skin, eyes, respiratory system, CNS
Strychnine	0.15 mg/m³	CNS
Styrene	100 ppm; 200 ppm ceiling; 600 ppm (5 min/3-hr peak)	CNS, respiratory system, eyes, skin
Sulfur dioxide	5 ppm (13 mg/m³)	Respiratory system, skin, eyes
Sulfuric acid	1 mg/m³	Respiratory system, eyes, skin, teeth
Sulfur mono-chloride	1 ppm (6 mg/m³)	Respiratory system, skin, eyes
Sulfur penta-fluoride	0.025 ppm (0.25 mg/m³)	Respiratory system, CNS
Sulfuryl fluoride	5 ppm (20 mg/m³)	Respiratory system, CNS
2,4,5-T	10 mg/m³	Skin, liver, GI tract
Talc (non-asbestiform)	20 mppcf	Lungs, CVS

National Occupational Standards   Continued

Chemical Name	Permissible Exposure Limit (8-hr TWA)*	Target Organ
Tantalum metal, oxide dusts (as Ta)	5 mg/m³	None known in humans
TEDP	0.2 mg/m³	CNS, respiratory system, CVS
Tellurium compounds (as Te)	0.1 mg/m³	Skin, CNS
Tellurium hexafluoride (as Te)	0.02 ppm (0.2 mg/m³)	Respiratory system
TEPP	0.05 mg/m³	CNS, respiratory system, CVS, GI tract
Terphenyls	1 ppm ceiling (9 mg/m³)	Skin, respiratory system
1,1,2,2-Tetrachloro-1,2-difluoroethane	500 ppm (4170 mg/m³)	Lungs, skin
1,1,1,2-Tetrachloro-2,2-difluoroethane	500 ppm (4170 mg/m³)	Respiratory system, skin
1,1,2,2-Tetrachloroethane	5 ppm (35 mg/m³)	Liver, kidneys, CNS
Tetrachloroethylene	100 ppm; 200 ppm ceiling; 300 ppm (5 min/3-hr peak)	Liver, kidneys, eyes, upper respiratory system, CNS
Tetrachloronaphthalene	2 mg/m³	Liver, skin
Tetraethyl lead (as Pb)	0.075 mg/m³	CNS, CVS, kidneys, eyes
Tetrahydrofuran	200 pm (590 mg/m³)	Eyes, skin, respiratory system, CNS
Tetramethyl lead	0.075 mg/m³	CNS, CVS, kidneys
Tetramethyl succinonitrile	0.5 ppm (3 mg/m³)	CNS
Tetranitromethane	1 ppm (8 mg/m³)	Respiratory system, eyes, skin, blood, CNS
Tetryl	1.5 mg/m³	Respiratory system, eyes, CNS, skin; in animals: liver, kidneys
Thallium, soluble compounds (as Tl)	0.1 mg/m³	Eyes, CNS, lung, liver, kidneys, GI tract, body hair
Thiram	5 mg/m³	Respiratory system, skin

National Occupational Standards   Continued

Chemical Name	Permissible Exposure Limit (8-hr TWA)*	Target Organ
Tin, inorganic compounds except oxides (as Sn)	2 mg/m³	Eyes, skin, respiratory system
Tin, organic compounds (as Sn)	0.1 mg/m³	CNS, eyes, liver, urinary tract, skin, blood
Titanium dioxide	15 mg/m³	Lungs
Toluene	200 ppm; 300 ppm ceiling; 500 ppm 10-min peak	CNS, liver, kidneys, skin
Toluene-2, 4-diisocyanate	0.02 ppm ceiling (0.14 mg/m³)	Respiratory system, skin
o-Toluidine	5 ppm (22 mg/m³)	Blood, kidneys, liver, CVS, skin, eyes
Tributyl phosphate	5 mg/m³	Respiratory system, skin, eyes
1,1,2-Trichloro-ethane	10 ppm (45 mg/m³)	CNS, eyes, nose, liver, kidneys
Trichloroethylene	100 ppm; 200 ppm ceiling; 300 ppm peak	Respiratory system, heart, liver, kidneys, CNS, skin
Trichloro-naphthalene	5 mg/m³	Skin, liver
1,2,3-Trichloro-propane	50 ppm (300 mg/m³)	Eyes, respiratory system, skin, CNS, liver
1,1,2-Trichloro-1,2,2,-trifluoro-ethane	1000 ppm (7600 mg/m³)	Skin, heart
Triethylamine	25 ppm (100 mg/m³)	Respiratory system, eyes, skin
Trifluoromono-bromomethane	1000 ppm (6100 mg/m³)	Heart, CNS
Trinitrotoluene	1.5 mg/m³	Blood, liver, eyes, CVS, CNS, kidneys, skin
Triorthocresyl phosphate	0.1 mg/m³	PNS, CNS
Triphenyl phosphate	3 mg/m³	Blood
Turpentine	100 ppm (560 mg/m³)	Skin, eyes, kidneys, respiratory system

National Occupational Standards   Continued

Chemical Name	Permissible Exposure Limit (8-hr TWA)*	Target Organ
Uranium, insoluble compounds (as U)	0.25 mg/m³	Skin, bone marrow, lymphatics
Uranium, soluble compounds (as U)	0.05 mg/m³	Respiratory system, blood, liver, lymphatics, kidneys, skin, bone marrow
Vanadium pentoxide dust (as V)	0.5 mg/m³ ceiling	Respiratory system, skin, eyes
Vanadium pentoxide fume (as V)	0.1 mg/m³ ceiling	Respiratory system, skin, eyes
Vinyl chloride	1 ppm; 5 ppm 15-min ceiling	Liver, CNS, blood, respiratory system, lymphatic system
Vinyltoluene	100 ppm (480 mg/m³)	Eyes, skin, respiratory system
Warfarin	0.1 mg/m³	Blood, CVS
Xylene (o-, m-, and p-isomers)	100 ppm (435 mg/m³)	CNS, eyes, GI tract, blood, liver, kidneys, skin
Xylidine	5 ppm (25 mg/m³)	Blood, lungs, liver, kidneys, CVS
Yttrium compounds (as Y)	1 mg/m³	Eyes, lungs
Zinc chloride fume	1 mg/m³	Respiratory system, skin, eyes
Zinc oxide fume	5 mg/m³	Respiratory system
Zirconium compounds (as Zr)	5 mg/m³	Respiratory system, skin

Regulations for OSHA-Designated Occupational Carcinogens

Chemical	Regulation
2-acetylaminofluorene alpha-naphthylamine 4-aminodiphenyl benzidine beta-napthylamine beta-propiolactone bis-chloromethyl ether 3,3-dichlorobenzidine 4-dimethylaminoazobenzene ethyleneimine methyl chloromethyl ether n-nitrosodimethylamine 4-nitrobiphenyl	Worker exposure is to be controlled through the required use of engineering controls, work practices, and personal protective equipment, including respirators

Adapted from NIOSH Pocket Guide to Chemical Hazards, September 1985.

Appendix D

GLOSSARY OF TOXICOLOGY TERMS

absorption the movement of a chemical into the bloodstream after its entrance into the body through the skin, lungs, or gastrointestinal tract

action level level or concentration of a chemical residue in food or feed above which adverse health effects are possible and above which corrective action should be taken

acute sharp, severe; having a *rapid onset*, severe symptoms and a relatively *short* course. In toxicology refers to a single large exposure to a chemical (*acute exposure*), or to the development of symptoms of poisoning *soon* after a *single exposure* to a substance (*acute toxicity*)

adsorption the process of attracting and holding other substances or particles to a surface. For example, skin may adsorb soil particles.

Ames assay test performed on bacteria to assess the capability of a chemical to cause mutations

cancer a disease characterized by malignant, uncontrolled growth of cells of body tissue

carcinogenesis bioassay study performed on experimental animals, usually rodents, to assess the potential of a chemical to cause cancer after lifetime, daily exposure

carcinogenicity/carcinogen the ability of a substance to cause cancer in a living organism/a substance capable of producing cancer in a living organism

case-control study epidemiological study that is performed by comparing the exposure histories of humans who show a particular toxic effect with normal individuals to investigate the possible causes of the toxicity

ceiling level   maximum allowable level or concentration of an airborne chemical in the workplace; not to be exceeded even instantaneously

cholinesterase   enzyme that is critical to the proper conduction of nerve impulses. Inhibition of this enzyme is a characteristic of the toxicity of some large classes of pesticides, e.g., organophosphates

chronic   occurring over a period of time. In toxicology refers to repeated exposure (*chronic* exposure) to a chemical for a long period of time or persistence of symptoms or disease over a long period of time

cohort study   epidemiological study that is performed by following a cohort of individuals into the future to elucidate factors which may cause toxicity

DDT   organochloride pesticide that was widely used in the United States for insect control and that is now banned. It is very persistent in the environment and can still be found in wildlife

Delaney amendment   clause in the Food, Drug and Cosmetic Act that bans the use of food additives that have been shown to induce cancer when ingested by man or animal

de minimus   legal doctrine that states that the law does not deal with insignificant things. It is applied in toxicology to chemicals that are of very low risk

dieldrin   organochlorine pesticide, in the same family as DDT, that was also widely used for insect control, is now banned, and is still found in the environment

distribution   movement of a substance from site of entry to other parts of the body

dose   a specified amount; a measure of exposure usually expressed as an amount per unit of body weight. For example, 2 mg per kg (2 mg/kg)

effect   the response produced due to a drug or chemical—*local effect* (an effect that occurs at the site of first contact); *systemic effect* (an effect that requires absorption and distribution of the substance and affects the body at a site distant to the entry point)

elimination   removal of a chemical from the body by metabolism or excretion

enzyme   protein molecule that acts as a catalyst in living organisms

enzyme induction   process by which enzyme levels are increased as a result of exposure of the organism to an inducer chemical

epidemiology   the study of the incidence, distribution, and control of disease in a population

excretion   removal of a substance or its metabolites from the body in urine, feces, perspiration, milk, or expired air

exposure   receiving a dose of a substance; contact with a chemical substance; *acute exposure* (a single, large dose); *chronic exposure* (receiving repeated doses over a period of time)

exposure assessment   measurement of the dose or amount of a chemical to which an individual has been exposed. May involve analyses of body fluids or environmental media

fetus   stage in the development of humans from the beginning of the third month of gestation up until birth

gastrointestinal   related to the stomach or intestines

genotoxic   describes a chemical that causes adverse effects in the genetic material of living organisms

hazard   the inherent adverse effect that a chemical poses

herbicide   a pesticide that kills plants

hydrogeology   study of the earth's surface with emphasis on the layers in which water is found

initiator   chemical that can cause the initial step in the process of carcinogenesis; generally thought to involve changes in the genetic material in affected cells

insecticide   a pesticide that kills insects

in vitro   describes studies that are done in the laboratory, literally in glass, as distinct from those performed using living animals

LD$_{50}$ the dose lethal to 50% of the animals being tested

MCL   maximum contaminant level, a legally enforceable maximum concentration allowable for a chemical in drinking water

metabolism   the biochemical changes that a chemical undergoes in the body

mutagen   a chemical that causes mutations, changes in the genetic material of an organism

mycotoxin   toxic chemical that is produced naturally by molds (fungi). A common mycotoxin is aflatoxin

NOAEL   the no observable adverse effect level, or the highest level at which a chemical causes no observable adverse effect in the species being tested

NOEL   the no observable effect level, or the highest level at which a chemical causes no observable changes in the species under investigation

nonpolar   describes a molecule that does not have a strong overall charge or polarity. Such molecules have limited solubility in water and greater solubility in lipids (fats)

organochlorine   describes a type of molecule made up of an organic core, basically hydrogen and carbon atoms, and chlorine side groups. DDT and dieldrin are examples of organochlorine compounds

organ toxicity   describes adverse effects that alter normal stucture, and/or functioning of specific organs; e.g., the liver

PEL   the permissible exposure limit, a maximum (legally enforceable) allowable level for a chemical in workplace air

pesticide   a chemical designed to kill specific, unwanted living things

placebo   describes a drug or treatment that has no known direct effect on disease but that works due to the patient's belief in its efficacy

polar   describes a molecule that has a strong overall charge or polarity. It is soluble in water but not in lipids (fats)

poison   a chemical substance harmful to living things

ppm (parts per million)   an expression describing a small concentration; an amount of substance in a million parts of another material. For example, 1 part (molecule) of salt in a million parts (molecules) of water

promoter   a chemical that, when administered after an initiator has been given, promotes the change of an initiated cell into a cancer

qualitative   referring to the occurrence of a substance without specifying its exact amount or concentration or describing an amount in general terms. For example, a large amount

quantitative   describing the amount of a substance in exact terms, for example, 27.5 pounds

receptor   a special molecule that recognizes and binds to a foreign chemical and is involved in the initial steps in a toxic response

renal   related to the kidney

reproductive toxicity   an effect that alters the normal reproductive functioning of an organism. An example of reproductive toxicity is loss of fertility

response   the reaction of the body to a chemical substance

risk   the probability that a substance will cause harm

risk assessment   the determination of the potential toxic effects due to chemical exposure in a particular situation. It involves use of both a toxicity assessment and an exposure assessment

risk management   the steps taken to reduce or eliminate the risk that has been revealed by the risk assessment

risk perception   the magnitude of the risk as it is perceived by an individual or population. This is a mixture of the measured risk and the preconceptions of the observer

safety   the opposite of hazard; the probability that injury will *not* result from a substance

safety factor   factor that is used to provide a margin of error when extrapolating from animal experimentation to estimate human risk. Commonly, the no-effect level in animals is divided by a safety factor of 100 to estimate a safe level in humans

standard   a general term used to describe legally established values above which sanctions will be applied. There are standards for workplace air (PELs) and drinking water (MCLs), for example

structure   the chemical and physical make-up of a substance

STEL   the short term exposure limit, the maximal allowable level in workplace air, usually measured over a fifteen-minute period

teratogen   an agent that can cause birth defects

TLV   the threshold limit value, the maximal allowable workplace air level for a chemical. It can be a time-weighted average (TLV-TWA), a short-term value (TLV-STEL), or an instantaneous value (TLV-Ceiling)

tolerance   level or concentration of a chemical residue in food or feed above which adverse health effects are possible and above which corrective action should be taken. (Similar to an action level, but established more formally so needs not be defended in court)

toxic   harmful; poisonous

toxicant   a poison

toxicity   the harmful effects produced by a poison

toxicity assessment   evaluation of the toxicity of a chemical based on all available human and animal data

toxicology   the study of the adverse effects of chemicals on living organisms

Bibliography

General Toxicology

Loomis, T.A. *Essentials of Toxicology*. Lea and Febiger, Philadelphia, PA, 1978.

Lu, F.C. *Basic Toxicology*. Hemisphere, Washington, DC, 1985.

Ottoboni, M.A. *The Dose Makes the Poison*. Vincente Books, Berkeley, CA, 1984.

Klaassen, C.D., Amdur, M.O., and Doull, J. (eds.). *Casarett and Doulls' Toxicology*. 3rd Edition, Macmillan Publishing Company, New York, NY, 1986.

Risk Assessment

The Conservation Foundation. *Risk Assessment and Risk Control*. The Conservation Foundation, Washington, DC, 1985

Clayson, D.B., Krewski, D., and Munro, I., (eds.). *Toxicological Risk Assessment*. CRC Press, Inc., Boca Raton, FL, 1985.

Lowrance, W.W. *Of Acceptable Risk*. William Kaufmann, Inc., Los Altos, CA, 1976.

Urquhart, J. and Heilmann, K. *Risk Watch*. Facts on File Publications, New York, NY, 1984.

Hallenbeck, W.H., and Cunningham, K.M. *Quantitative Risk Assessment for Environmental and Occupational Health*. Lewis Publishers, Inc., Chelsea, MI, 1986.

Shrader-Frechette, K.S. *Risk Analysis and Scientific Method*. D. Reidel Publishing Co., Boston, MA, 1985.

Chemical Manufacturers Association. *Risk Analysis in the Chemical Industry*. Government Institutes, Inc., Rockville, MD, 1985.

Nelkin, D. *The Language of Risk*. Sage Publications, Beverly Hills, CA, 1985.

National Academy of Sciences. *Risk Assessment in the Federal Government: Managing the Process*. National Academy Press, Washington, DC, 1983.

Kopfler, F. and Craun, G. *Environmental Epidemiology*. Lewis Publishers, Inc., Chelsea, MI, 1986.

Artificial Sweeteners

National Academy of Sciences. *Evaluation of Cyclamate for Carcinogenicity*. National Academy Press, Washington, DC, 1985.

Office of Technology Assessment. *Cancer Testing Technology and Saccharin*. U.S. Government Printing Office, Washington, DC, 1985.

Turner, J.S. *The Chemical Feast*. Grossman Publishers, New York, NY, 1970

Whalen, E.M., and Stare, F.J. *Panic in the Pantry*. Atheneum, New York, NY, 1975.

Formaldehyde

Occupational Safety and Health Administration. Occupational Exposure to Formaldehyde. *Federal Register* 50: 50412–50499, 1985.

Report on the Consensus Workshop on Formaldehyde. *Env. Health Pers.* 58: 323–381, 1984.

Asbestos

National Academy of Sciences. *Asbestiform Fibers: Nonoccupational Health Risks*. National Academy Press, Washington, DC, 1984.

Ontario (Canada) Ministry of the Attorney General. *Report of the Royal Commission on Matters of Health and Safety Arising from the Use of Asbestos in Ontario*. Ontario Ministry of Government Services, Toronto, Canada, 1984.

Benzene

Occupational Safety and Health Administration. Occupational Exposure to Benzene. *Federal Register* 50: 50512–50586, 1985.

Environmental Toxicants

D'Itri, F.M., and Kamrin, M. (eds.). *PCBs: Human and Environmental Hazards*. Ann Arbor Science Publishers, Ann Arbor, MI, 1983.

Kamrin, M., and Rodgers, P. (eds.). *Dioxins in the Environment*. Hemisphere Publishing Corp., New York, NY, 1985.

Gammage, R.B., and Kaye, S.V. *Indoor Air and Human Health*. Lewis Publishers, Inc., Chelsea, MI, 1985.

Turiel, I. *Indoor Air Quality and Human Health*. Stanford University Press, Stanford, CA, 1985.

Rappe, C., Choudhary, G., and Keith, L.H. *Chlorinated Dioxins and Dibenzofurans in Perspective*. Lewis Publishers, Inc., Chelsea, MI, 1986.

National Academy of Sciences. *Drinking Water and Health, Vol. 1–6*. National Academy Press, Washington, DC, 1977–1986.

Index

143

DATE DUE	BORROWER'S NAME	ROOM NUMBER